1
arbook of Astronomy

1981 Yearbook of Astronomy

PATRICK MOORE

W · W · Norton & Company
NEW YORK, LONDON

54284

Printed in Great Britain

Contents

PART TWO: ARTICLE SECTION

PART THREE: MISCELLANEOUS

Editor's Foreword

This nineteenth edition of the *Yearbook* follows along the same lines as before. In the Article section, emphasis has been placed upon subjects of special current interest, and among our new contributors we welcome John Mason, Dr Leonard Culhane, Gordon Taylor, and James E. Oberg of the United States; they join our familiar and ever-welcome authors – Dr Garry Hunt, Dr David Allen, Professor Arnold Wolfendale, and Martin Cohen.

There is, however, one sad note. Ever since the *Yearbook* began, in 1962 (how long ago that seems!), Dr J.G. Porter has contributed all the charts and notes for Part One, which is a vital section of the publication. He has also been of tremendous help in all other ways. He has now, alas, decided that the time has come for him to retire, and therefore his charts and notes in the present *Yearbook* will be his last.

Everyone will regret this, but we hope that Dr Porter will continue to make contributions whenever he feels able to do so. Meanwhile, we thank him very sincerely for all his work in the past, and send him our best wishes for the future.

PATRICK MOORE

Selsey, March 1980

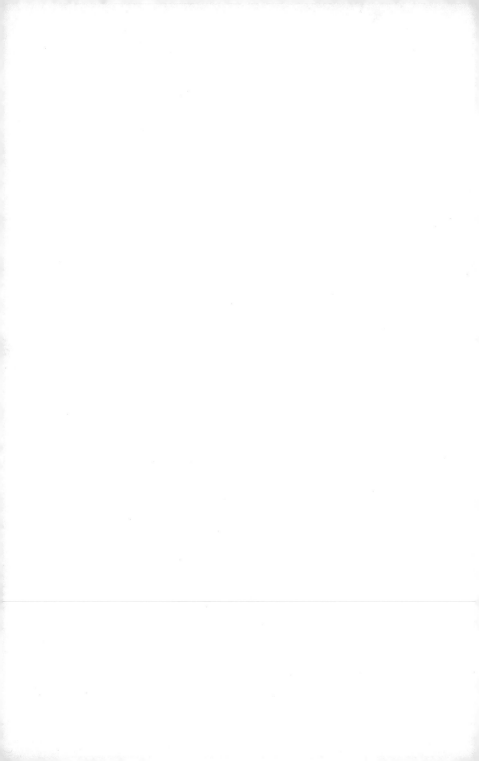

Preface

New readers will find that all the information in this *Yearbook* is given in diagrammatic or descriptive form; the positions of the planets may easily be found on the specially designed star charts, while the monthly notes describe the movements of the planets and give details of other astronomical phenomena visible in both the northern and southern hemispheres. Two sets of star charts are provided. The **Northern Charts** (pp. 16 to 41) are designed for use in latitude 52 degrees north, but may be used without alteration throughout the British Isles, and (except in the case of eclipses and occultations) in other countries of similar north latitude. The **Southern Charts** (pp. 42 to 67) are drawn for latitude 35 degrees south, and are suitable for use in South Africa, Australia and New Zealand, and other stations in approximately the same south latitude. The reader who needs more detailed information will find *Norton's Star Atlas* (Gall and Inglis) an invaluable guide, while more precise positions of the planets and their satellites, together with predictions of occultations, meteor showers, and periodic comets may be found in the *Handbook* of the British Astronomical Association. A somewhat similar publication is the *Observer's Handbook* of the Royal Astronomical Society of Canada, and readers will also find details of forthcoming events given in the American *Sky and Telescope*. This monthly publication also produces a special occultation supplement giving predictions for the United States and Canada.

Important Note
    The times given on the star charts and in the Monthly Notes are generally given as local times, using the 24-hour clock, the day beginning at midnight. The times of a few events (e.g.

eclipses) are given in Greenwich Mean Time (G.M.T.), which is
related to local time by the formula

Local Mean Time = G.M.T. — west longitude.

In practice, small differences of longitude are ignored, and the
observer will use local clock time, which will be the appropriate
Standard (or Zone) Time. As the formula indicates, places in
west longitude will have a standard Time slow on G.M.T.,
while places in east longitude will have Standard Times fast on
G.M.T. As examples we have:

Standard Time in

New Zealand	G.M.T.	+	12 hours
Victoria; N.S.W.	G.M.T.	+	10 hours
Western Australia	G.M.T.	+	8 hours
South Africa	G.M.T.	+	2 hours
British Isles	G.M.T.		
Eastern S.T.	G.M.T.	–	5 hours
Central S.T.	G.M.T.	–	6 hours, etc.

If Summer Time is in use, the clocks will have been advanced
by one hour, and this hour must be subtracted from the clock
time to give Standard Time.

In Great Britain and N. Ireland, Summer Time will be in
force in 1981 from 29 March 02^h until 25 October 02^h G.M.T.

PART ONE

Events of 1981

MONTHLY CHARTS and
ASTRONOMICAL PHENOMENA

Notes on the Star Charts

The stars, together with the Sun, Moon and planets seem to be set on the surface of the celestial sphere, which appears to rotate about the Earth from east to west. Since it is impossible to represent a curved surface accurately on a plane, any kind of star map is bound to contain some form of distortion. But it is well known that the eye can endure some kinds of distortion better than others, and it is particularly true that the eye is most sensitive to deviations from the vertical and horizontal. For this reason the star charts given in this volume have been designed to give a true representation of vertical and horizontal lines, whatever may be the resulting distortion in the shape of a constellation figure. It will be found that the amount of distortion is, in general, quite small, and is only obvious in the case of large constellations such as Leo and Pegasus, when these appear at the top of the charts, and so are drawn out sideways.

The charts show all stars down to the fourth magnitude, together with a number of fainter stars which are necessary to define the shape of a constellation. There is no standard system for representing the outlines of the constellations, and triangles and other simple figures have been used to give outlines which are easy to follow with the naked eye. The names of the constellations are given, together with the proper names of the brighter stars. The apparent magnitudes of the stars are indicated roughly by using four different sizes of dots, the larger dots representing the bright stars.

The two sets of star charts are similar in design. At each opening there is a group of four charts which give a complete coverage of the sky up to an altitude of 62½ degrees; there are twelve such groups to cover the entire year. In the **Northern Charts** (for 52 degrees north) the upper two charts show the southern sky, south being at the centre and east on the left. The coverage is from 10 degrees north of east (top left) to 10 degrees north of west (top right). The two lower charts show the northern sky from 10 degrees south of west (lower left) to 10 degrees south of east (lower right). There is thus an overlap east and west.

Conversely, in the **Southern Charts** (for 35 degrees south) the upper two charts show the northern sky, with north at the centre and east on the right. The two lower charts show the southern sky, with south at the centre and east on the left. The coverage and overlap is the same on both sets of charts.

Because the sidereal day is shorter than the solar day, the stars appear to rise and set about four minutes earlier each day, and this amounts to two hours in a month. Hence the twelve groups of charts in each set are sufficient to give the appearance of the sky throughout the day at intervals of two hours, or at the same time of night at monthly intervals throughout the year. The actual range of dates and times when the stars on the charts are visible is indicated at the top of each page. Each group is numbered in bold type, and the number to be used for any given month and time is summarised in the following table:

Local Time	18h	20h	22h	0h	2h	4h	6h
January	11	12	1	2	3	4	5
February	12	1	2	3	4	5	6
March	1	2	3	4	5	6	7
April	2	3	4	5	6	7	8
May	3	4	5	6	7	8	9
June	4	5	6	7	8	9	10
July	5	6	7	8	9	10	11
August	6	7	8	9	10	11	12
September	7	8	9	10	11	12	1
October	8	9	10	11	12	1	2
November	9	10	11	12	1	2	3
December	10	11	12	1	2	3	4

The charts are drawn to scale, the horizontal measurements, marked at every 10 degrees, giving the azimuths (or true bearings) measured from the north round through east (90 degrees), south (180 degrees), and west (270 degrees). The vertical measurements, similarly marked, give the altitudes of the stars up to 62½ degrees. Estimates of altitude and azimuth made from these charts will necessarily be mere approximations, since no observer will be exactly at the adopted latitude, or at the stated time, but they will serve for the identification of stars and planets.

The ecliptic is drawn as a broken line on which longitude is marked at every 10 degrees; the positions of the planets are then easily found by reference to the table on page 74. It will be noticed that on the southern charts the ecliptic may reach an altitude in excess of 62½ degrees on star charts 5 to 9. The continuation of the broken line will be found on the charts of overhead stars.

There is a curious illusion that stars at an altitude of 60 degrees or more are actually overhead, and the beginner may often feel that he is leaning over backwards in trying to see them. These overhead stars are given separately on the pages immediately following the main star charts. The entire year is covered at one opening, each of the four maps showing the overhead stars at times which correspond to those of three of the main star charts. The position of the zenith is indicated by a cross, and this cross marks the centre of a circle which is 35 degrees from the zenith; there is thus a small overlap with the main charts.

The broken line leading from the north (on the Northern Charts) or from the south (on the Southern Charts) is numbered to indicate the corresponding main chart. Thus on page 40 the N-S line numbered 6 is to be regarded as an extension of the centre (south) line of chart 6 on pages 26 and 27, and at the top of these pages are printed the dates and times which are appropriate. Similarly, on page 67, the S-N line numbered 10 connects with the north line of the upper charts on page 60.

The overhead stars are plotted as maps on a conical projection, and the scale is rather smaller than that of the main charts.

1L

October 6 at 5^h	October 21 at 4^h
November 6 at 3^h	November 21 at 2^h
December 6 at 1^h	December 21 at midnight
January 6 at 23^h	January 21 at 22^h
February 6 at 21^h	February 21 at 20^h

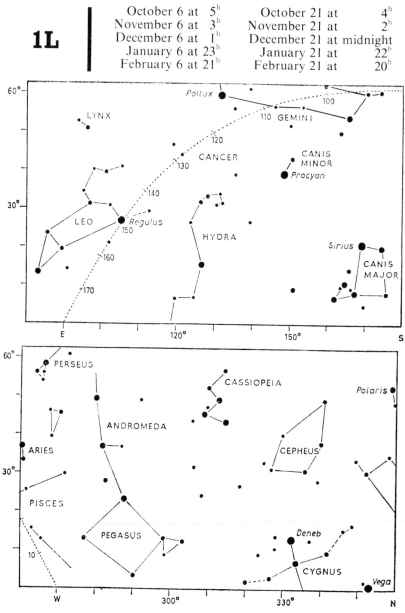

October 6 at 5^h October 21 at 4^h
November 6 at 3^h November 21 at 2^h
December 6 at 1^h December 21 at midnight
January 6 at 23^h January 21 at 22^h
February 6 at 21^h February 21 at 20^h

1R

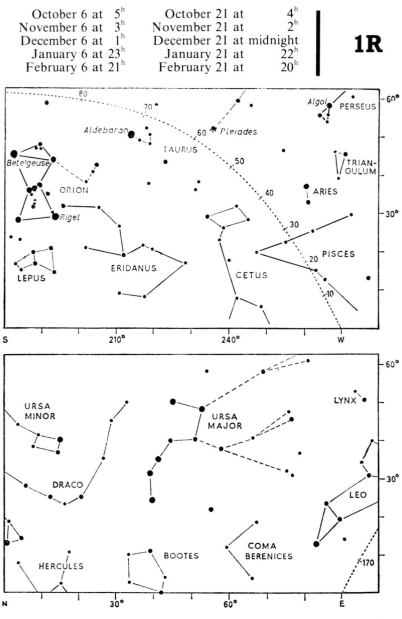

17

2L

November 6 at 5^h	November 21 at 4^h
December 6 at 3^h	December 21 at 2^h
January 6 at 1^h	January 21 at midnight
February 6 at 23^h	February 21 at 22^h
March 6 at 21^h	March 21 at 20^h

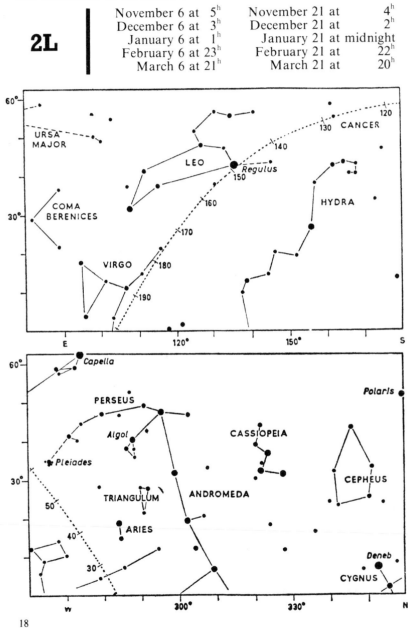

November 6 at 5^h November 21 at   4^h
December 6 at 3^h December 21 at   2^h
January 6 at 1^h January 21 at midnight
February 6 at 23^h February 21 at   22^h
March 6 at 21^h March 21 at   20^h

2R

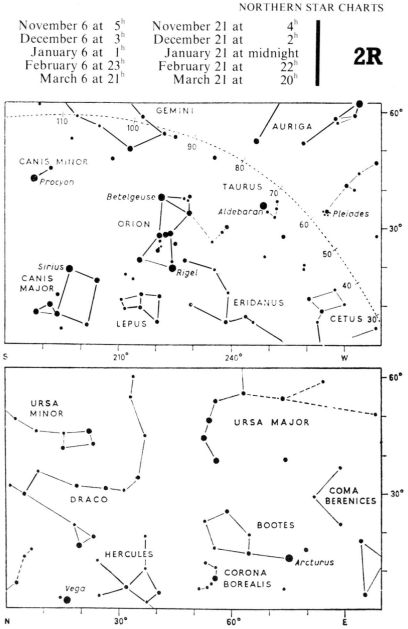

3L

December 6 at 5^h	December 21 at 4^h
January 6 at 3^h	January 21 at 2^h
February 6 at 1^h	February 21 at midnight
March 6 at 23^h	March 21 at 22^h
April 6 at 21^h	April 21 at 20^h

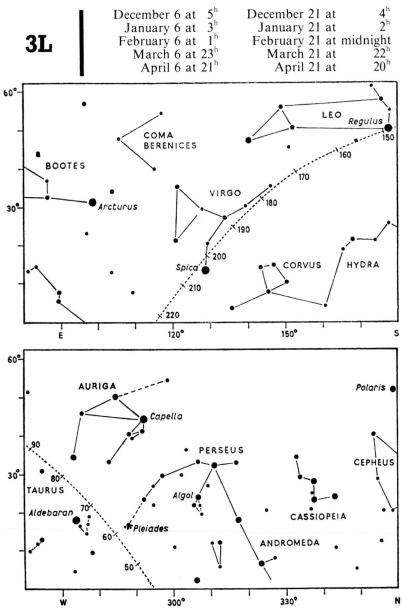

December 6 at 5^h	December 21 at 4^h
January 6 at 3^h	January 21 at 2^h
February 6 at 1^h	February 21 at midnight
March 6 at 23^h	March 21 at 22^h
April 6 at 21^h	April 21 at 20^h

3R

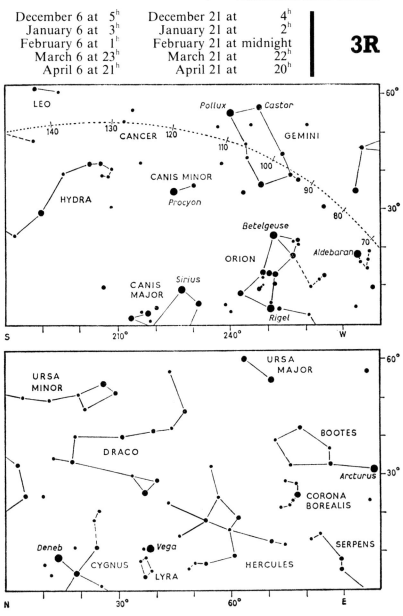

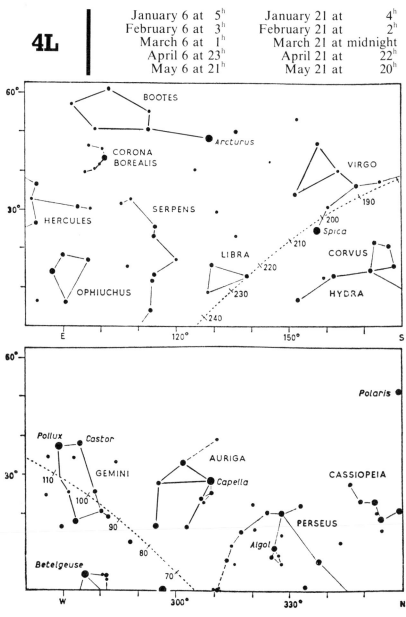

January 6 at 5^h	January 21 at 4^h
February 6 at 3^h	February 21 at 2^h
March 6 at 1^h	March 21 at midnight
April 6 at 23^h	April 21 at 22^h
May 6 at 21^h	May 21 at 20^h

4L

January 6 at 5^h	January 21 at 4^h
February 6 at 3^h	February 21 at 2^h
March 6 at 1^h	March 21 at midnight
April 6 at 23^h	April 21 at 22^h
May 6 at 21^h	May 21 at 20^h

4R

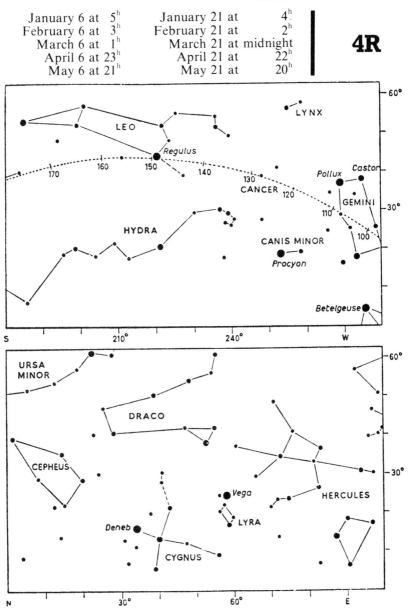

5L

January 6 at 7^h	January 21 at 6^h
February 6 at 5^h	February 21 at 4^h
March 6 at 3^h	March 21 at 2^h
April 6 at 1^h	April 21 at midnight
May 6 at 23^h	May 21 at 22^h

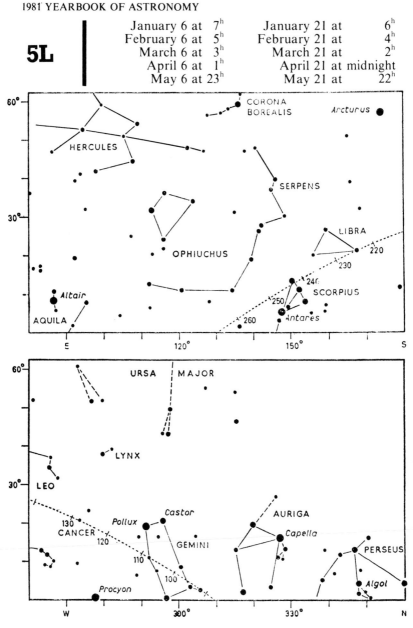

January 6 at 7^h	January 21 at 6^h
February 6 at 5^h	February 21 at 4^h
March 6 at 3^h	March 21 at 2^h
April 6 at 1^h	April 21 at midnight
May 6 at 23^h	May 21 at 22^h

5R

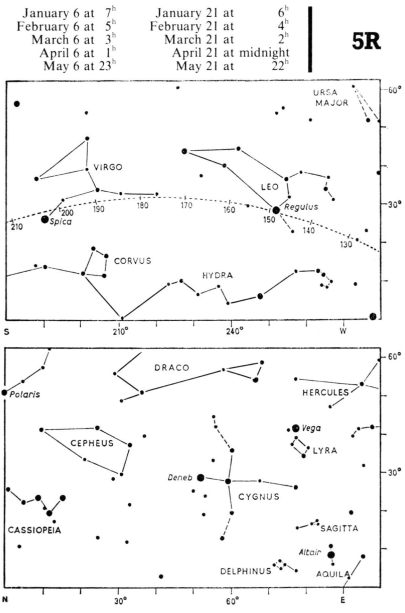

6L

March 6 at 5^h	March 21 at 4^h
April 6 at 3^h	April 21 at 2^h
May 6 at 1^h	May 21 at midnight
June 6 at 23^h	June 21 at 22^h
July 6 at 21^h	July 21 at 20^h

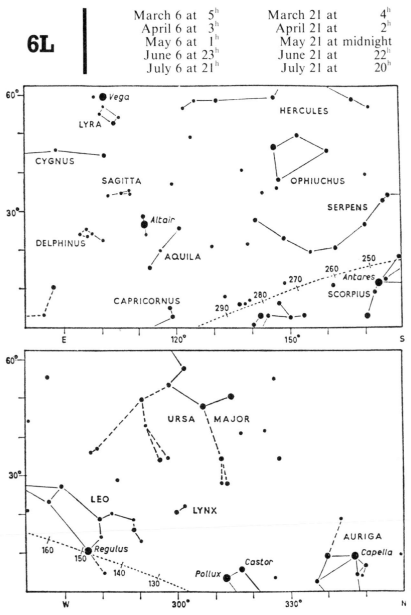

March 6 at 5^h March 21 at 4^h
April 6 at 3^h April 21 at 2^h
May 6 at 1^h May 21 at midnight
June 6 at 23^h June 21 at 22^h
July 6 at 21^h July 21 at 20^h

6R

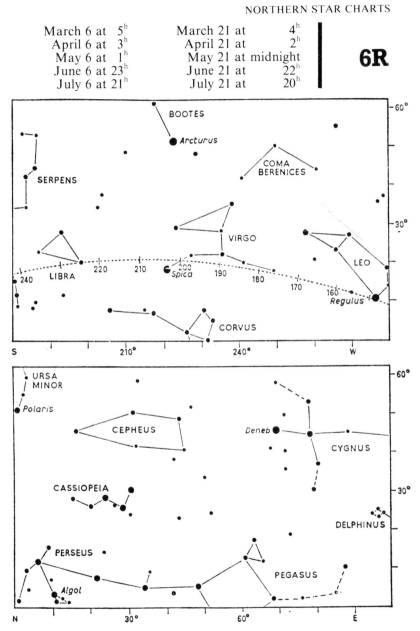

7L

May 6 at 3^h	May 21 at 2^h
June 6 at 1^h	June 21 at midnight
July 6 at 23^h	July 21 at 22^h
August 6 at 21^h	August 21 at 20^h
September 6 at 19^h	September 21 at 18^h

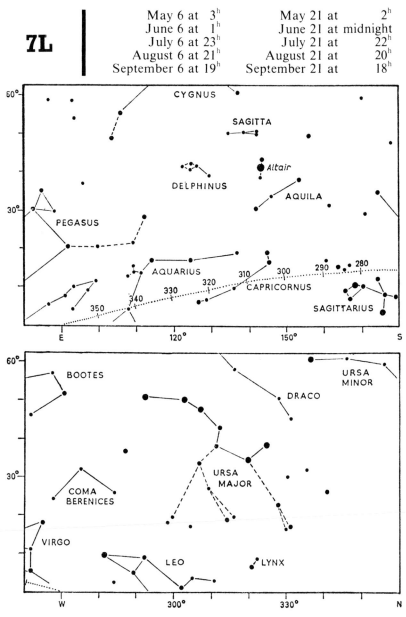

May 6 at 3^h	May 21 at 2^h
June 6 at 1^h	June 21 at midnight
July 6 at 23^h	July 21 at 22^h
August 6 at 21^h	August 21 at 20^h
September 6 at 19^h	September 21 at 18^h

7R

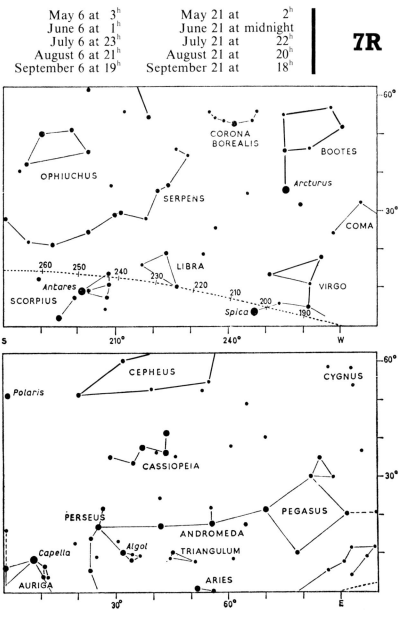

8L

July 6 at 1ʰ
August 6 at 23ʰ
September 6 at 21ʰ
October 6 at 19ʰ
November 6 at 17ʰ

July 21 at midnight
August 21 at 22ʰ
September 21 at 20ʰ
October 21 at 18ʰ
November 21 at 16ʰ

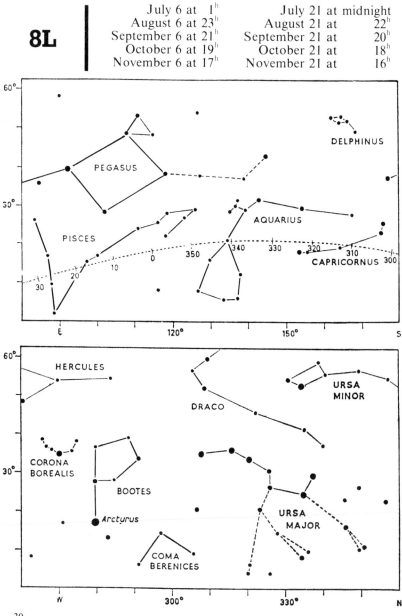

July 6 at 1^{h} July 21 at midnight
August 6 at 23^{h} August 21 at 22^{h}
September 6 at 21^{h} September 21 at 20^{h}
October 6 at 19^{h} October 21 at 18^{h}
November 6 at 17^{h} November 21 at 16^{h}

8R

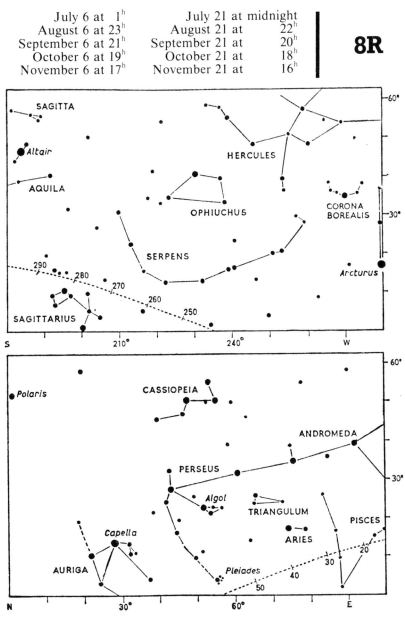

9L

August 6 at 1^h	August 21 at midnight
September 6 at 23^h	September 21 at 22^h
October 6 at 21^h	October 21 at 20^h
November 6 at 19^h	November 21 at 18^h
December 6 at 17^h	December 21 at 16^h

August 6 at 1^h August 21 at midnight
September 6 at 23^h September 21 at   22^h
October 6 at 21^h October 21 at   20^h
November 6 at 19^h November 21 at   18^h
December 6 at 17^h December 21 at   16^h

9R

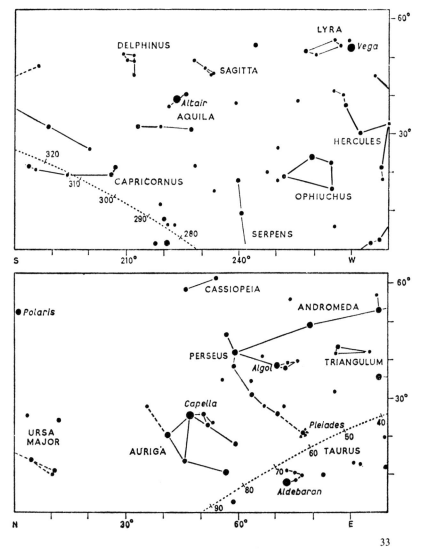

33

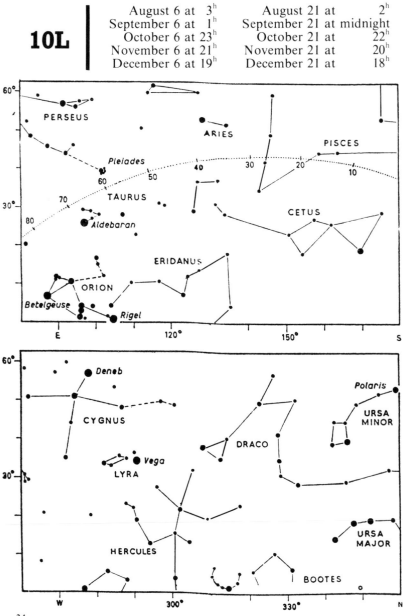

August 6 at 3^h August 21 at 2^h
September 6 at 1^h September 21 at midnight
October 6 at 23^h October 21 at 22^h
November 6 at 21^h November 21 at 20^h
December 6 at 19^h December 21 at 18^h

10R

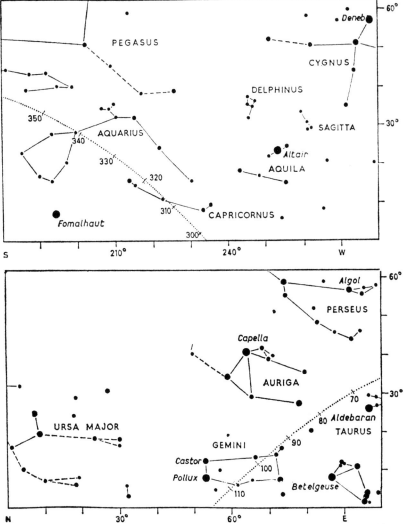

September 6 at 3^h	September 21 at 2^h
October 6 at 1^h	October 21 at midnight
November 6 at 23^h	November 21 at 22^h
December 6 at 21^h	December 21 at 20^h
January 6 at 19^h	January 21 at 18^h

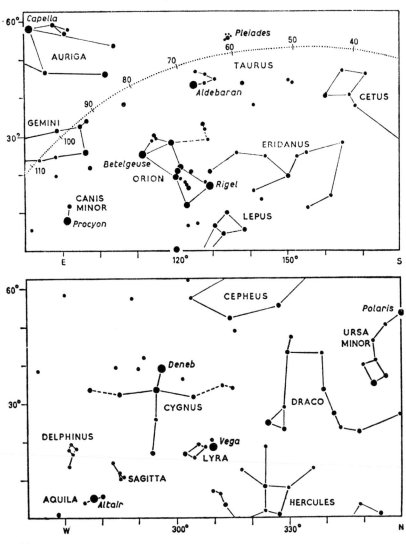

September 6 at 3ʰ	September 21 at 2ʰ
October 6 at 1ʰ	October 21 at midnight
November 6 at 23ʰ	November 21 at 22ʰ
December 6 at 21ʰ	December 21 at 20ʰ
January 6 at 19ʰ	January 21 at 18ʰ

11R

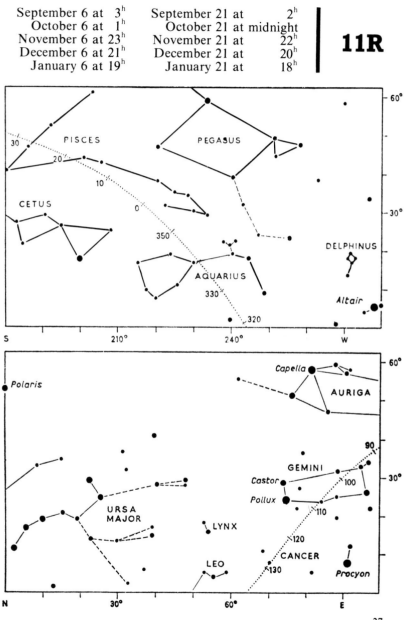

12L

October 6 at 3ʰ | October 21 at 2ʰ
November 6 at 1ʰ | November 21 at midnight
December 6 at 23ʰ | December 21 at 22ʰ
January 6 at 21ʰ | January 21 at 20ʰ
February 6 at 19ʰ | February 21 at 18ʰ

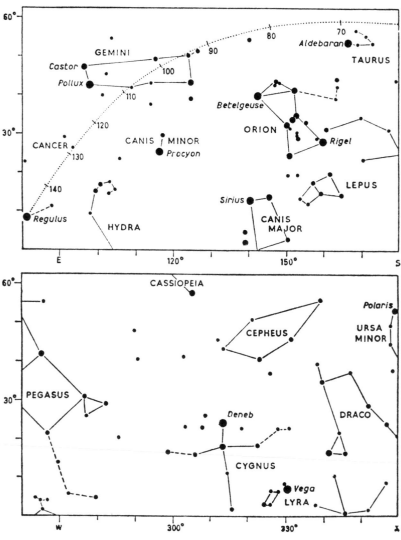

October 6 at 3^h	October 21 at 2^h
November 6 at 1^h	November 21 at midnight
December 6 at 23^h	December 21 at 22^h
January 6 at 21^h	January 21 at 20^h
February 6 at 19^h	February 21 at 18^h

12R

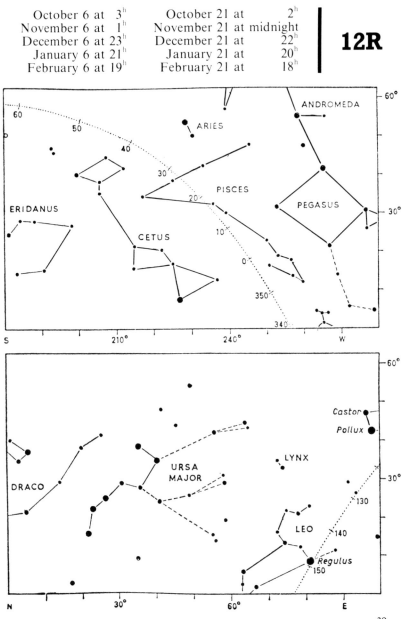

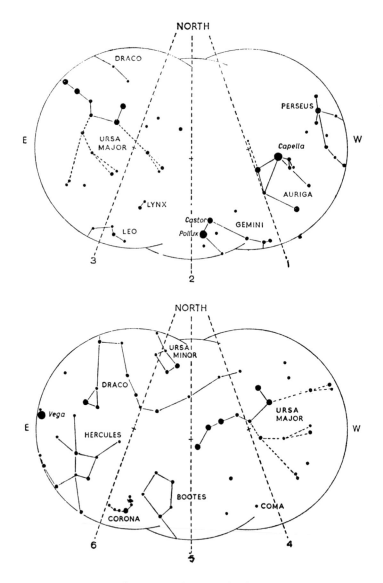

Northern Hemisphere Overhead Stars

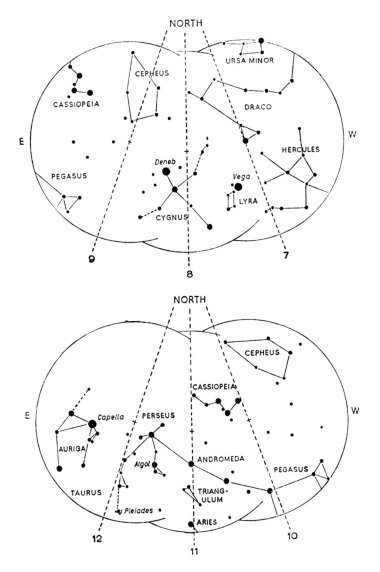

Northern Hemisphere Overhead Stars

1L

October 6 at 5^h	October 21 at 4^h
November 6 at 3^h	November 21 at 2^h
December 6 at 1^h	December 21 at midnight
January 6 at 23^h	January 21 at 22^h
February 6 at 21^h	February 21 at 20^h

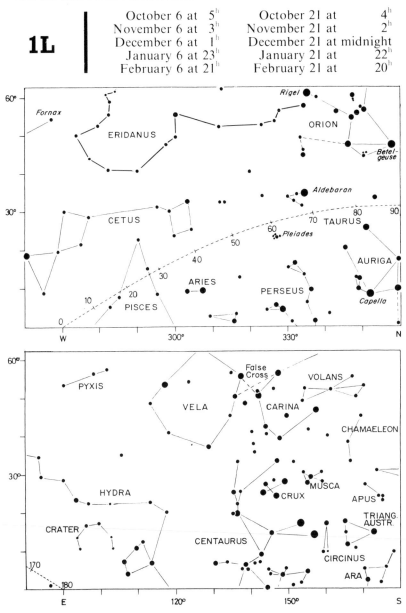

October 6 at 5^h	October 21 at 4^h	
November 6 at 3^h	November 21 at 2^h	**1R**
December 6 at 1^h	December 21 at midnight	
January 6 at 23^h	January 21 at 22^h	
February 6 at 21^h	February 21 at 20^h	

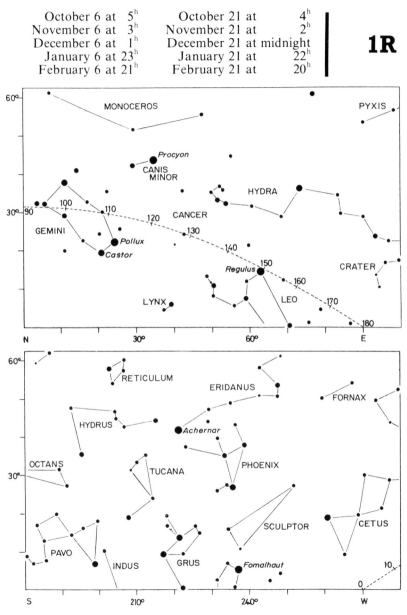

2L

November 6 at 5^h	November 21 at 4^h
December 6 at 3^h	December 21 at 2^h
January 6 at 1^h	January 21 at midnight
February 6 at 23^h	February 21 at 22^h
March 6 at 21^h	March 21 at 20^h

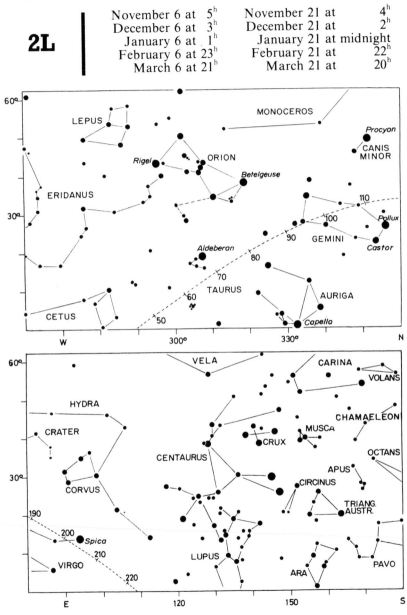

November 6 at 5^h	November 21 at	4^h
December 6 at 3^h	December 21 at	2^h
January 6 at 1^h	January 21 at midnight	
February 6 at 23^h	February 21 at	22^h
March 6 at 21^h	March 21 at	20^h

2R

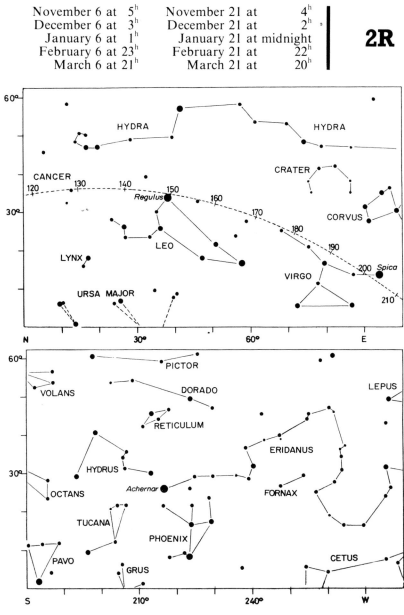

3L

January 6 at   3ʰ   January 21 at   2ʰ
February 6 at   1ʰ   February 21 at midnight
March 6 at 23ʰ   March 21 at   22ʰ
April 6 at 21ʰ   April 21 at   20ʰ
May 6 at 19ʰ   May 21 at   18ʰ

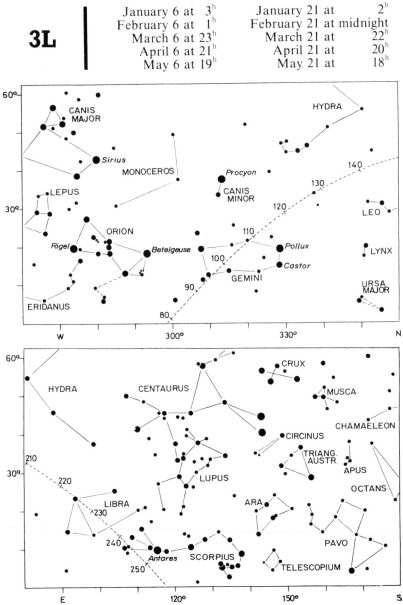

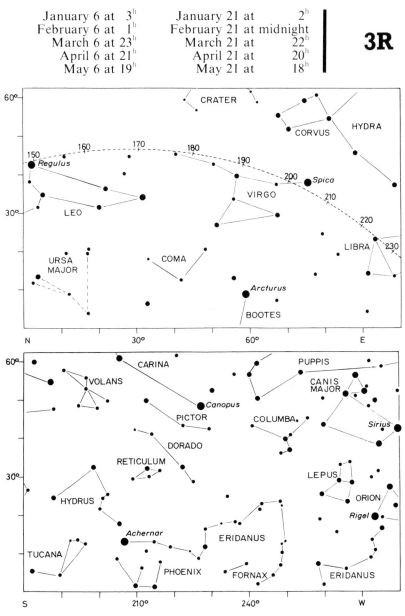

January 6 at 3^h January 21 at 2^h
February 6 at 1^h February 21 at midnight
March 6 at 23^h March 21 at 22^h
April 6 at 21^h April 21 at 20^h
May 6 at 19^h May 21 at 18^h

3R

CRATER
CORVUS
HYDRA
150 160 170 180 190 200 210 220 230
Regulus
Spica
VIRGO
LIBRA
LEO
COMA
URSA
MAJOR
Arcturus
BOOTES
60°
30°
N 30° 60° E

CARINA
PUPPIS
VOLANS
CANIS
MAJOR
Canopus
PICTOR
COLUMBA
Sirius
DORADO
RETICULUM
LEPUS
ORION
HYDRUS
Rigel
Achernar
ERIDANUS
TUCANA
PHOENIX
FORNAX
ERIDANUS
60°
30°
S 210° 240° W

4L

February 6 at 3ʰ	February 21 at 2ʰ
March 6 at 1ʰ	March 21 at midnight
April 6 at 23ʰ	April 21 at 22ʰ
May 6 at 21ʰ	May 21 at 20ʰ
June 6 at 19ʰ	June 21 at 18ʰ

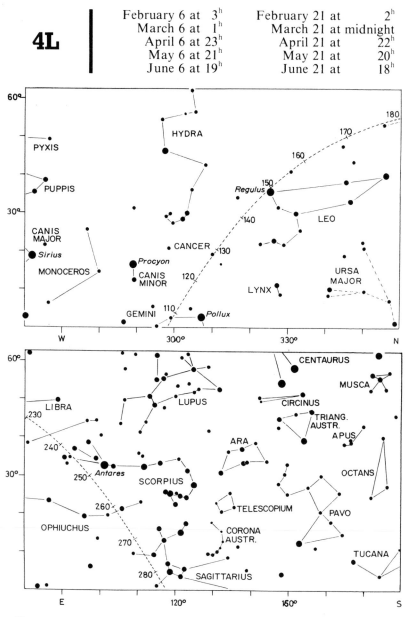

February 6 at 3ʰ   February 21 at 2ʰ
March 6 at 1ʰ   March 21 at midnight
April 6 at 23ʰ   April 21 at 22ʰ
May 6 at 21ʰ   May 21 at 20ʰ
June 6 at 19ʰ   June 21 at 18ʰ

4R

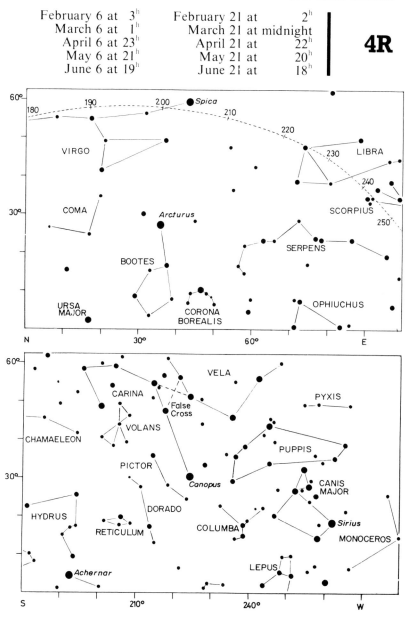

5L

March 6 at 3^h	March 21 at 2^h
April 6 at 1^h	April 21 at midnight
May 6 at 23^h	May 21 at 22^h
June 6 at 21^h	June 21 at 20^h
July 6 at 19^h	July 21 at 18^h

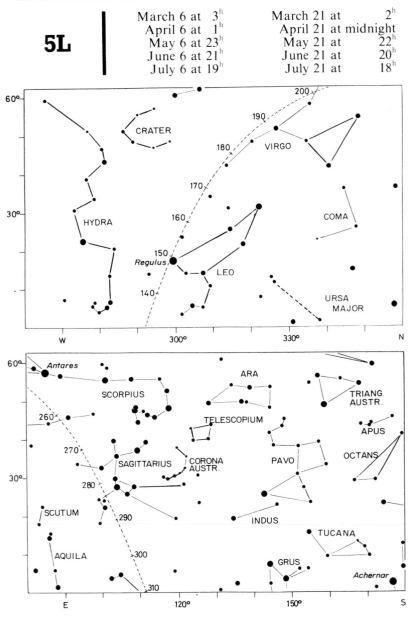

March 6 at   3ʰ   March 21 at   2ʰ
April 6 at   1ʰ   April 21 at midnight
May 6 at 23ʰ   May 21 at   22ʰ   **5R**
June 6 at 21ʰ   June 21 at   20ʰ
July 6 at 19ʰ   July 21 at   18ʰ

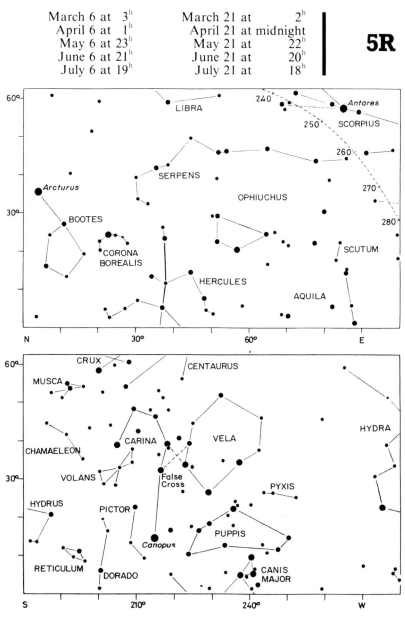

6L

March 6 at 5^h	March 21 at 4^h
April 6 at 3^h	April 21 at 2^h
May 6 at 1^h	May 21 at midnight
June 6 at 23^h	June 21 at 22^h
July 6 at 21^h	July 21 at 20^h

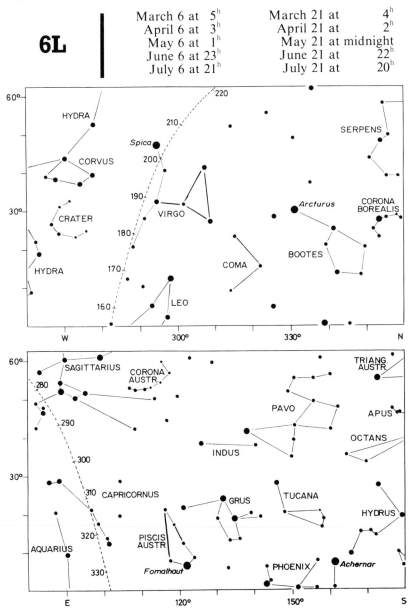

March 6 at 5^h March 21 at 4^h
April 6 at 3^h April 21 at 2^h
May 6 at 1^h May 21 at midnight
June 6 at 23^h June 21 at 22^h
July 6 at 21^h July 21 at 20^h

6R

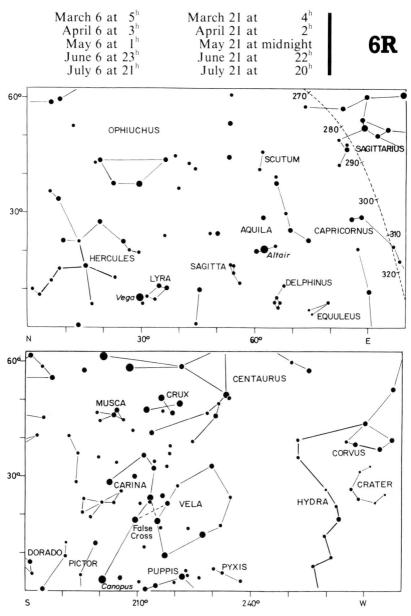

7L

April 6 at 5^h	April 21 at 4^h
May 6 at 3^h	May 21 at 2^h
June 6 at 1^h	June 21 at midnight
July 6 at 23^h	July 21 at 22^h
August 6 at 21^h	August 21 at 20^h

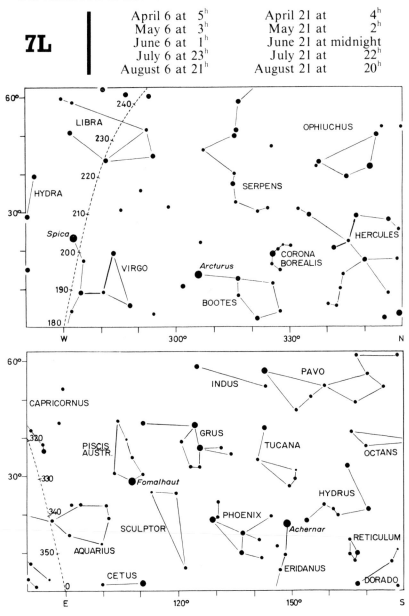

April 6 at 5ʰ   April 21 at   4ʰ
May 6 at 3ʰ   May 21 at   2ʰ
June 6 at 1ʰ   June 21 at midnight
July 6 at 23ʰ   July 21 at   22ʰ
August 6 at 21ʰ   August 21 at   20ʰ

7R

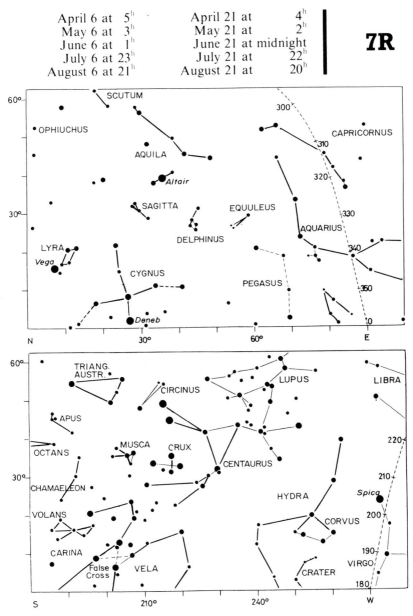

8L

May 6 at 5^h

June 6 at 3^h

July 6 at 1^h

August 6 at 23^h

September 6 at 21^h

May 21 at 4^h

June 21 at 2^h

July 21 at midnight

August 21 at 22^h

September 21 at 20^h

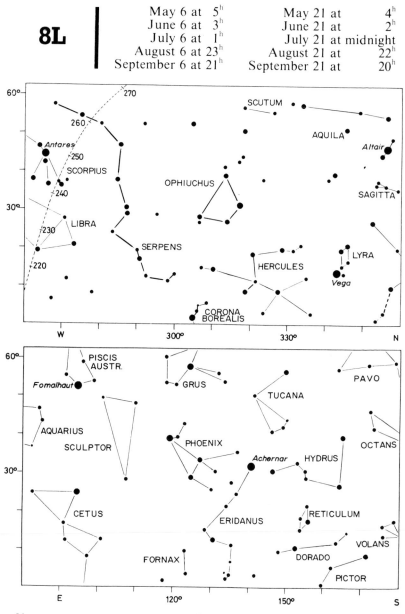

May 6 at   5ʰ   May 21 at   4ʰ
June 6 at   3ʰ   June 21 at   2ʰ
July 6 at   1ʰ   July 21 at midnight
August 6 at 23ʰ   August 21 at   22ʰ
September 6 at 21ʰ   September 21 at   20ʰ

8R

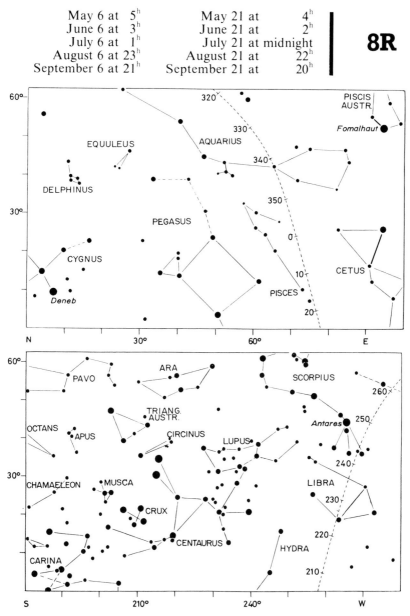

9L

June 6 at 5ʰ	June 21 at 4ʰ
July 6 at 3ʰ	July 21 at 2h
August 6 at 1ʰ	August 21 at midnight
September 6 at 23ʰ	September 21 at 22ʰ
October 6 at 21ʰ	October 21 at 20ʰ

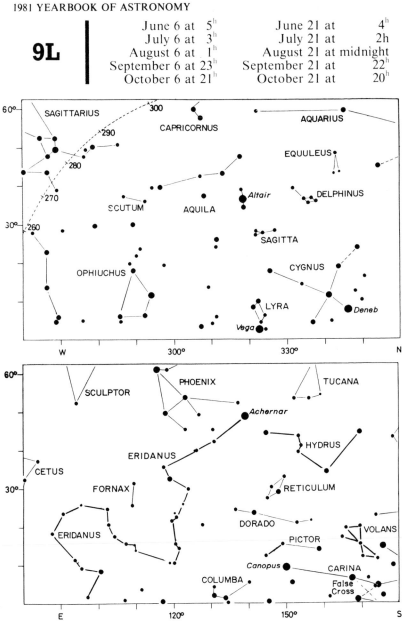

June 6 at 5^h June 21 at   4^h
July 6 at 3^h July 21 at   2h
August 6 at 1^h August 21 at midnight
September 6 at 23^h September 21 at   22^h
October 6 at 21^h October 21 at   20^h

9R

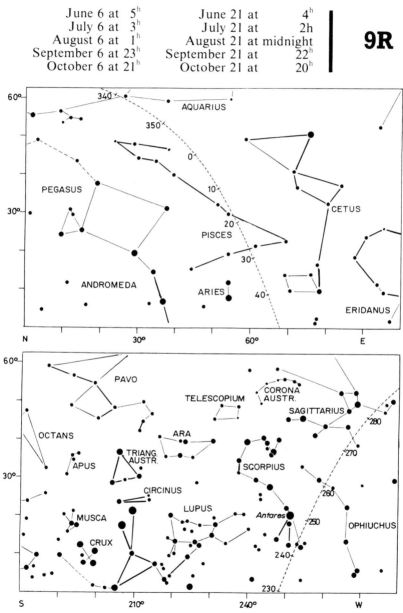

10L

July 6 at 5^h	July 21 at 4^h
August 6 at 3^h	August 21 at 2^h
September 6 at 1^h	September 21 at midnight
October 6 at 23^h	October 21 at 22^h
November 6 at 21^h	November 21 at 20^h

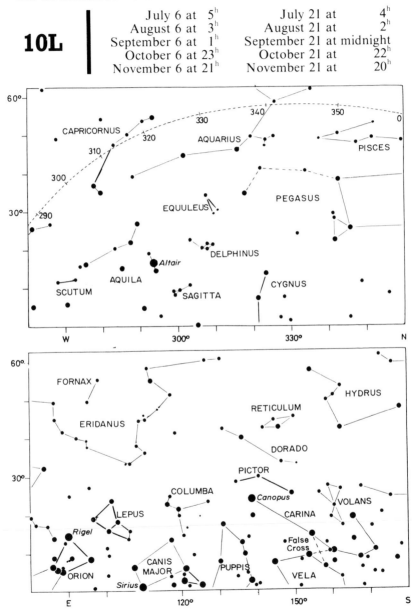

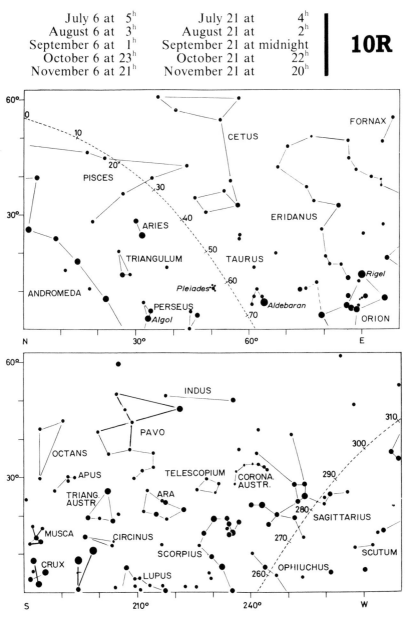

11L

August 6 at 5ʰ	August 21 at 4ʰ
September 6 at 3ʰ	September 21 at 2ʰ
October 6 at 1ʰ	October 21 at midnight
November 6 at 23ʰ	November 21 at 22ʰ
December 6 at 21ʰ	December 21 at 20ʰ

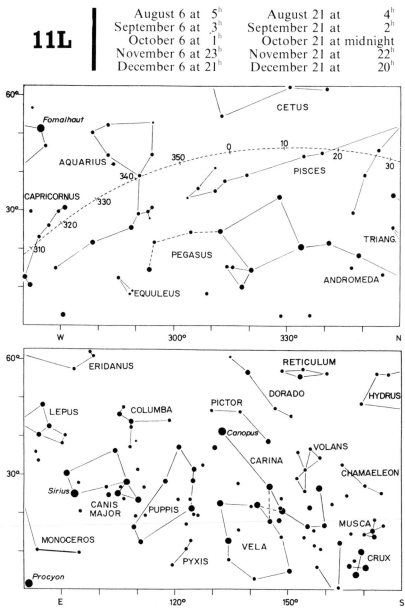

August 6 at 5^h August 21 at   4^h
September 6 at 3^h September 21 at   2^h **11R**
October 6 at 1^h October 21 at midnight
November 6 at 23^h November 21 at   22^h
December 6 at 21^h December 21 at   20^h

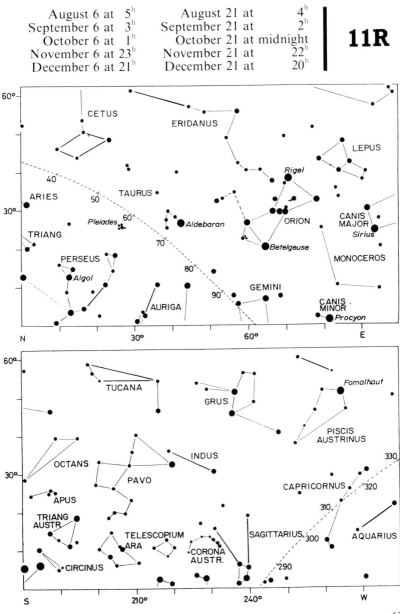

12L

September 6 at 5^h	September 21 at 4^h
October 6 at 3^h	October 21 at 2^h
November 6 at 1^h	November 21 at midnight
December 6 at 23^h	December 21 at 22^h
January 6 at 21^h	January 21 at 20^h

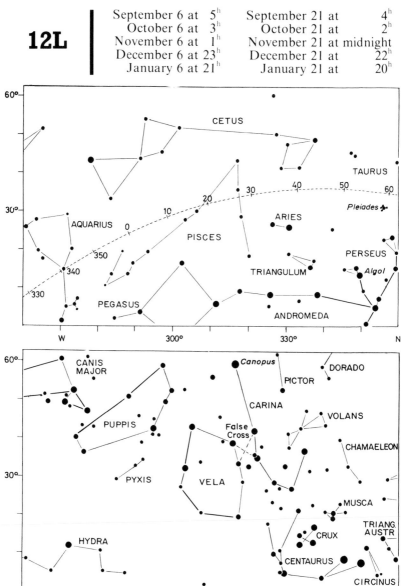

September 6 at 5ʰ   September 21 at   4ʰ
October 6 at   3ʰ   October 21 at   2ʰ
November 6 at 1ʰ   November 21 at midnight
December 6 at 23ʰ   December 21 at   22ʰ
January 6 at 21ʰ   January 21 at   20ʰ

12R

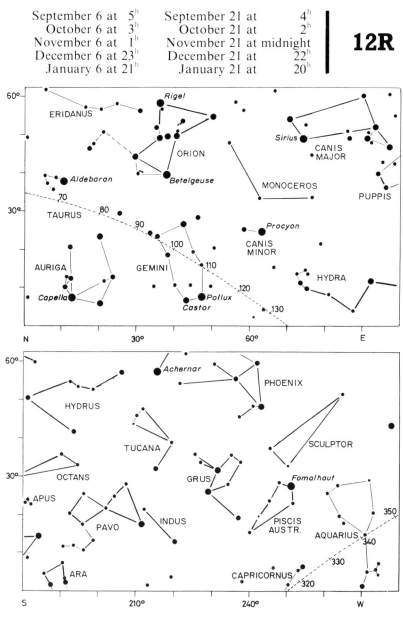

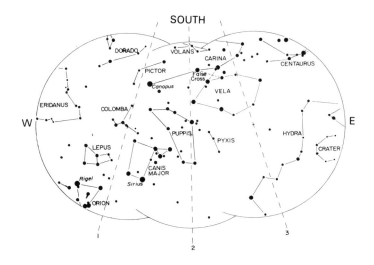

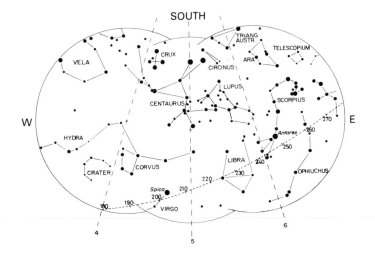

Southern Hemisphere Overhead Stars

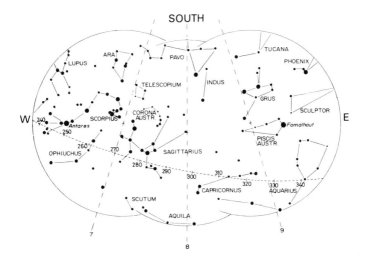

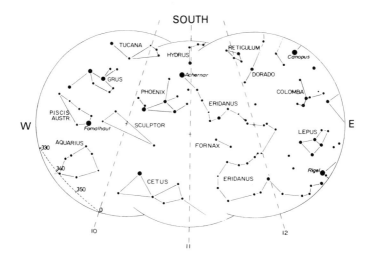

Southern Hemisphere Overhead Stars

The Planets and the Ecliptic

The paths of the planets about the Sun all lie close to the plane of the ecliptic, which is marked for us in the sky by the apparent path of the Sun among the stars, and is shown on the star charts by a broken line. The Moon and planets will always be found close to this line, never departing from it by more than about 7 degrees. Thus the planets are most favourably placed for observation when the ecliptic is well displayed, and this means that it should be as high in the sky as possible. This avoids the difficulty of finding a clear horizon, and also overcomes the problem of atmospheric absorption, which greatly reduces the light of the stars. Thus a star at an altitude of 10 degrees suffers a loss of 60 per cent of its light, which corresponds to a whole magnitude; at an altitude of only 4 degrees, the loss may amount to two magnitudes.

The position of the ecliptic in the sky is therefore of great importance, and since it is tilted at about 23½ degrees to the equator, it is only at certain times of the day or year that it is displayed to the best advantage. It will be realized that the Sun (and therefore the ecliptic) is at its highest in the sky at noon in midsummer, and at its lowest at noon in midwinter. Allowing for the daily motion of the sky, these times lead to the fact that the ecliptic is highest at midnight in winter, at sunset in the spring, at noon in summer and at sunrise in the autumn. Hence these are the best times to see the planets. Thus, if Venus is an evening star, in the western sky after sunset, it will be seen to best advantage if this occurs in the spring, when the ecliptic is high in the sky and slopes down steeply to the north-west. This means that the planet is not only higher in the sky, but will remain for a much longer period above the horizon. For similar reasons, a morning star will be seen at its best on autumn mornings before sunrise, when the ecliptic is high in the east. The outer planets, which can come to opposition (i.e. opposite the Sun), are best seen when opposition occurs in the winter months, when the ecliptic is high in the sky at midnight.

The seasons are reversed in the Southern Hemisphere, spring beginning at the September Equinox, when the Sun crosses the Equator on its way south, summer begins at the December Solstice, when the Sun is highest in the southern sky, and so on. Thus, the times when the ecliptic is highest in the sky, and therefore best placed for observing the planets, may be summarised as follows:

	Midnight	*Sunrise*	*Noon*	*Sunset*
Northern lats.	December	September	June	March
Southern lats.	June	March	December	September

In addition to the daily rotation of the celestial sphere from east to west, the planets have a motion of their own among the stars. The apparent movement is generally *direct,* i.e. to the east, in the direction of increasing longitude, but for a certain period (which depends on the distance of the planet) this apparent motion is reversed. With the outer planets this *retrograde* motion occurs about the time of opposition. Owing to the different inclination of the orbits of these planets, the actual effect is to cause the apparent path to form a loop, or sometimes an S-shaped curve. The same effect is present in the motion of the inferior planets, Mercury and Venus, but it is not so obvious, since it always occurs at the time of inferior conjunction.

The inferior planets, Mercury and Venus, move in smaller orbits than that of the Earth, and so are always seen near the Sun. They are most obvious at the times of greatest angular distance from the Sun (greatest elongation), which may reach 28 degrees for Mercury, or 47 degrees for Venus. They are then seen as evening stars in the western sky after sunset (at eastern elongations) or as morning stars in the eastern sky before sunrise (at western elongations). The succession of phenomena, conjunctions and elongations, always follows the same order, but the intervals between them are not equal. Thus, if either planet is moving round the far side of its orbit its motion will be to the east, in the same direction in which the Sun appears to be moving. It therefore takes much longer for the planet to overtake the Sun—that is, to come to superior conjunction—than it does when moving round to inferior conjunction, between Sun and Earth. The intervals given in the following table are average values; they remain fairly constant in the case of Venus, which travels in an almost circular orbit. In the case of Mercury, however, conditions vary widely because of the great eccentricity and inclination of the planet's orbit.

		Mercury	Venus
Inferior conj.	to Elongation West	22 days	72 days
Elongation West	to Superior conj.	36 days	220 days
Superior conj.	to Elongation East	36 days	220 days
Elongation East	to Inferior conj.	22 days	72 days

The greatest brilliancy of Venus always occurs about 36 days before or after inferior conjunction. This will be about a month *after* greatest eastern elongation (as an evening star), or a month *before* greatest western elongation (as a morning star). No such rule can be given for Mercury, because its distance from the Earth and the Sun can vary over a wide range.

Mercury is not likely to be seen unless a clear horizon is available. It is seldom seen as much as 10 degrees above the horizon in the twilight sky in northern latitudes, but this figure is often exceeded in the Southern Hemisphere. This favourable condition arises because the maximum elongation of 28 degrees can occur only when the planet is at aphelion (farthest from the Sun), and this point lies well south of the Equator. Northern observers must be content with smaller elongations, which may be as little as 18 degrees at perihelion. In general, it may be said that the most favourable times for seeing Mercury as an evening star will be in spring, some days before greatest eastern elongation; in autumn, it may be seen as a morning star some days after greatest western elongation.

Venus is the brightest of the planets, and may be seen on occasions in broad daylight. Like Mercury, it is alternately a morning and an evening star, and will be highest in the sky when it is a morning star in autumn, or an evening star in spring. The phenomena of Venus given in the table above can occur only in the months of January, April, June, August and November, and it will be realized that they do not all lead to favourable apparitions of the planet. In fact, Venus is to be seen at its best as an evening star in northern latitudes when eastern elongation occurs in June. The planet is then well north of the Sun in the preceding spring months, and is a brilliant object in the evening sky over a long period. In the Southern Hemisphere a November elongation is best. For similar reasons, Venus gives a prolonged display as a morning star in the months following western elongation in November (in northern latitudes) or in June (in the Southern Hemisphere).

The superior planets, which travel in orbits larger than that of the Earth, differ from Mercury and Venus in that they can be seen opposite the Sun in the sky. The superior planets are morning stars after conjunction with the Sun, rising earlier each day until they come to opposition. They will then be nearest to the Earth (and therefore at their brightest), and will be on the meridian at midnight, due south in northern latitudes, but due north in the Southern Hemisphere. After opposition they are evening stars, setting earlier each evening until they set in the west with the Sun at the next conjunction. The change in brightness about the time of opposition is most noticeable in the case of Mars, whose distance from the Earth can vary considerably and rapidly. The other superior planets are at such great distances that there is very little change in brightness from one opposition to another. The effect of altitude is, however, of some importance, for at a December opposition in northern latitudes the planet will be among the stars of Taurus or Gemini, and can then be at an altitude of more than 60 degrees in southern England. At a summer opposition, when the planet is in Sagittarius, it may only rise to about 15 degrees above the southern horizon, and so makes a less impressive appearance. In the Southern Hemisphere, the reverse conditions apply; a June opposition being the best, with the planet in Sagittarius at an altitude which can reach 78 degrees above the northern horizon.

Mars, whose orbit is appreciably eccentric, comes nearest to the Earth at an opposition at the end of August. It may then be brighter even than Jupiter, but rather low in the sky in Aquarius for northern observers, though very well placed for those in southern latitudes. These favourable oppositions occur every fifteen or seventeen years (1924, 1941, 1956, 1971) but in the Northern Hemisphere the planet is probably better seen at an opposition in the autumn or winter months, when it is higher in the sky. Oppositions of Mars occur at an average interval of 780 days, and during this time the planet makes a complete circuit of the sky.

Jupiter is always a bright planet, and comes to opposition a month later each year, having moved, roughly speaking, from one Zodiacal constellation to the next.

Saturn moves much more slowly than Jupiter, and may remain in the same constellation for several years. The brightness of Saturn depends on the aspect of its rings, as well as on the distance from Earth and Sun. The rings are now inclined towards the Earth and Sun at quite a small angle, and are opening again after being seen edge-on in 1980.

Uranus, Neptune, and *Pluto* are hardly likely to attract the attention of observers without adequate instruments, but some notes on their present positions in the sky will be found in the April, May and June Notes.

Phases of the Moon 1981

New Moon	First Quarter	Full Moon	Last Quarter
d h m	d h m	d h m	d h m
Jan. 6 07 24	Jan. 13 10 10	Jan. 20 07 39	Jan. 28 04 19
Feb. 4 22 14	Feb. 11 17 49	Feb. 18 22 58	Feb. 27 01 14
Mar. 6 10 31	Mar.13 01 50	Mar. 20 15 22	Mar. 28 19 34
Apr. 4 20 19	Apr. 11 11 11	Apr. 19 07 59	Apr. 27 10 14
May 4 04 19	May 10 22 22	May 19 00 04	May 26 21 00
June 2 11 32	June 9 11 33	June 17 15 04	June 25 04 25
July 1 19 03	July 9 02 39	July 17 04 39	July 24 09 40
July 31 03 52	Aug. 7 19 26	Aug. 15 16 37	Aug. 22 14 16
Aug. 29 14 43	Sept. 6 13 26	Sept. 14 03 09	Sept. 20 19 47
Sept. 28 04 07	Oct. 6 07 45	Oct. 13 12 49	Oct. 20 03 40
Oct. 27 20 13	Nov. 5 01 09	Nov. 11 22 26	Nov. 18 14 54
Nov. 26 14 38	Dec. 4 16 22	Dec. 11 08 41	Dec. 18 05 47
Dec. 26 10 10			

All times are G.M.T.

Reproduced, with permission, from data supplied by the Science Research Council

The Planets in 1981

DATE		Venus	Mars	Jupiter	Saturn	Uranus	Neptune
January	6	263	305	190	190	239	263
	21	282	317	190	190	239	264
February	6	302	329	190	189	240	264
	21	321	341	189	189	240	264
March	6	337	351	188	188	240	265
	21	356	3	186	187	240	265
April	6	16	15	184	186	240	265
	21	34	27	182	184	239	265
May	6	53	38	181	184	239	264
	21	71	49	180	183	238	264
June	6	90	61	181	183	237	264
	21	109	71	181	183	237	263
July	6	127	82	183	184	236	263
	21	146	92	184	185	236	263
August	6	165	102	187	186	236	262
	21	183	112	190	187	236	262
September	6	202	122	193	189	236	262
	21	219	132	196	191	237	262
October	6	237	141	199	193	238	262
	21	253	150	202	195	239	263
November	6	270	159	206	197	239	263
	21	285	167	209	198	240	264
December	6	298	175	212	200	241	264
	21	307	182	214	201	242	265
Opposition:		——	——	Mar. 26	Mar. 27	May 19	June 14
Conjunction:		Apr. 7	Apr. 2	Oct. 14	Oct. 6	Nov. 22	Dec. 16

Mercury moves so quickly among the stars that it is not possible to indicate its position on the star charts at a convenient interval. The monthly notes must be consulted for the best times at which the planet may be seen.

The positions of the other planets are given in the table on the previous page. This gives the apparent longitudes on dates which correspond to those of the star charts, and the position of the planet may at once be found near the ecliptic at the given longitude.

Examples:

(1) *Where may the planets Mars and Saturn be found in the first week of December?*

The table opposite gives the longitudes of these planets as Mars 175 degrees and Saturn 200 degrees. For northern latitudes Chart 3L shows these positions rising south of east at 5^h, just before dawn. Mars is below the figure of Leo, while Saturn is lower in the sky above the star Spica. In the summer skies of the Southern Hemisphere conditions are less favourable, and Southern Chart 2R shows the eastern sky at the earlier time of 3^h. Mars is then above the inverted figure of Leo, while Saturn is due east to the north of Spica.

(2) *In the Southern Hemisphere two bright planets are seen setting in the west soon after dark at the end of August. Identify both planets.*

Southern Chart 7L shows the western sky at 20^h on 21 August, and the two planets must have longitudes about 180 degrees to 190 degrees. The table shows Venus at 183 degrees, Jupiter at 187 degrees and Saturn at 190 degrees, but Venus and Jupiter are always far brighter than Saturn, and their positions relative to the star Spica will also confirm their identity. Northern observers will use Northern Chart 7R, which shows that these planets are very low in the western sky at this time, but it may be possible to see Venus and Jupiter in the bright evening sky before they set.

Some Events in 1981

ECLIPSES

In 1981 there will be three eclipses, two of the Sun, and one of the Moon.

4-5 February – an annular eclipse of the Sun, visible in Australasia and the west of South America.

17 July – a partial eclipse of the Moon, visible in Europe, Africa and America.

31 July – a total eclipse of the Sun, visible in eastern Europe, most of Asia and the north-west of North America.

THE PLANETS

Mercury may best be seen in northern latitudes at eastern elongation (evening star) on 27 May, and at western elongation (morning star) on 3 November. In the Southern Hemisphere conditions are very favourable on 16 March (morning star) and 23 September (evening star).

Venus will continue as a morning star until superior conjunction in April. After this it will be an evening star, reaching greatest eastern elongation on 11 November.

Mars is in superior conjunction on 2 April, and there is no opposition of Mars in 1981.

Jupiter is at opposition on 26 March in Virgo. Conjunction is on 14 October.

Saturn is at opposition on 27 March, also in Virgo, and conjunction is on 6 October.

Uranus is at opposition on 19 May in Libra.

Neptune is at opposition on 14 June in Ophiuchus.

Pluto is at opposition on 13 April on the borders of Virgo and Boötes.

MONTHLY NOTES 1981

January

New Moon: 6 January   *Full Moon:* 20 January

Earth is at perihelion (nearest to the Sun) on 2 January at a distance of 91.4 million miles (147.1 million km).

Mercury was in superior conjunction on 31 December, but by the end of January it will be approaching eastern elongation. It is then quite bright and may be seen in the south-west after sunset. Mars is in the same part of the sky, and Mercury passes 0.3 degrees south of Mars on the night of 23 January, but Mars will be difficult to see in the bright twilight sky. Mercury is not so well placed in southern latitudes, and is only a few degrees above the horizon at sunset at the end of the month.

Venus is a morning star, rising in the south-east about an hour before the Sun at the beginning of the year. Magnitude –3.4 to –3.3. In northern latitudes Venus closes in to the Sun quite rapidly, but it is well south of the Equator and will be seen over a longer period in the Southern Hemisphere.

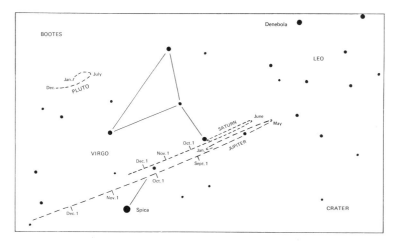

Mars is an evening star in Capricornus for a short time at the beginning of the year. At the end of the month it passes into Aquarius, setting about an hour after the Sun, but it is not very bright (magnitude +1.4) and will be difficult to find in a bright sky. There is no opposition of Mars in 1981, and there will be no marked increase in the brightness of this planet until the end of the year.

Jupiter is a morning star, moving direct in Virgo and rising in the east shortly after midnight. Since the planet is now near the Equator, the time of rising is approximately the same in all latitudes. The direct motion of Jupiter continues until 25 January, when it reaches a stationary point near the third magnitude double star Gamma Virginis. Before this happens, Jupiter will pass rather more than a degree south of Saturn. The actual conjunction occurs on 14 January, and this is the first of three conjunctions of these two planets at this opposition, the others occurring in February and July. This triple conjunction is quite rare (see comments in the February notes), and can only occur when the times of the oppositions of the two planets are very close. Magnitude of Jupiter –1.6 to –1.8.

Saturn rises close to Jupiter shortly after midnight, and is also moving direct in Virgo until 19 January when it reaches a stationary point near Gamma Virginis. The conjunction with Jupiter on 14 January will emphasize the contrast in brightness of the two planets. The magnitude of Saturn is $+1.0$ to $+0.9$, so that Jupiter is more than ten times as bright. The diagram shows the paths of both planets during the year, but the loops in the paths at the time of opposition are shown on a larger scale in the diagram in the March notes. The rings of Saturn are now opening again, the north side being visible and illuminated by the Sun.

A penumbral eclipse of the Moon on 20 January will be of little general interest. It may be observed in its entirety in North and South America. For details, see page 124.

CERES AND THE ALBEDOES OF THE ASTEROIDS

Ceres reaches opposition this month; it is in Gemini, not far from Castor and Pollux. It is of magnitude 6.5, so that it is not far below naked-eye visibility, and can be seen with binoculars, though it looks exactly like a star, and may be identified only by its movement from one night to the next.

Ceres is much the largest of the asteroids, and has more than twice the diameter of any other apart from Pallas (unless we include Chiron, whose orbit lies mainly beyond that of Saturn, and which cannot be ranked as an ordinary asteroid). It moves round the Sun at a distance ranging between 2.55 and 2.94 astronomical units, in a period of 4.06 years, so that it is in the main asteroid zone; its distance corresponds very nearly to that required by Bode's Law (though since this alleged Law breaks down for Neptune, and is not exact for most of the other planets, it may well be nothing more than a happy coincidence; opinions differ). The rotation period is known to be 9.03 hours. In view of its size, Ceres is probably spherical rather than irregular in shape, as with the smaller members of the minor planet swarm.

Ceres is not the brightest of the asteroids. This honour belongs to Vesta, which is smaller, but has a higher albedo or reflecting power. Asteroid diameters have been under survey lately, as described by Gordon Taylor elsewhere in this *Yearbook;* so too are the albedoes, and there is a considerable range. Values currently accepted for some of the senior asteroids are as follows:

Asteroid	Albedo
1 Ceres	0.054
2 Pallas	0.074
3 Juno	0.151
4 Vesta	0.229
5 Astræa	0.140
10 Hygeia	0.041
18 Melpomene	0.144

The most reflective asteroid seems to be 44 Nysa, with an albedo of 0.377. At the other end of the scale comes 95 Arethusa, with 0.019. This means that Arethusa is blacker than a blackboard; it has wrested the title from the former holder, 342 Bamberga, with 0.032.

As yet we know nothing definite about the surface details of asteroids, and we must await space-probe results. However, it may be that Phobos and Deimos, the dwarf satellites of Mars, are ex-asteroids; if their surfaces are typical of the swarm, then we may expect to find that most or all of the asteroids are cratered.

COMET REINMUTH 2

Reinmuth's second periodical comet is due at perihelion around January 30. It was discovered in 1947; the period is 6.75 years, and as it has been observed at all its subsequent returns its orbit is well known. This year it is in the Southern Hemisphere of the sky, in Capricornus, but as the magnitude is expected to be no brighter than 19 it is beyond the range of most amateur observers.

TWO ASTRONOMICAL ANNIVERSARIES

January 12 is the anniversary of the death of a noted Italian astronomer, Ercole Dembowski. He was born in Milan in

1812, and established a private observatory near Naples, later transferring it to Milan. He was a specialist in the observation of double stars, and, with the Struves, was outstanding in pioneer double-star measurement.

This is also the centenary of the birth of Francis Gledheim Pease. He was born at Cambridge, Massachusetts, on 14 January 1881, and educated in Illinois; he went to the Yerkes Observatory in 1901, and to Mount Wilson in 1904. He collaborated with Harlow Shapley in studies of stellar clusters, and in 1920 designed an interferometer with which he made the first direct measurements of the diameter of a star (Betelgeux). He obtained fine lunar pictures from Mount Wilson, and was a noted telescope designer. Pease died on 7 February 1938.

February

New Moon: 4 February   *Full Moon:* 18 February

Mercury is at greatest eastern elongation (18 degrees) on 2 February and will be visible in the early days of the month as an evening star in the south-west after sunset. It is not so well placed in the Southern Hemisphere, and is not likely to be seen. Mercury is in inferior conjunction on 17 February and towards the end of the month it is a morning star in the east before sunrise. It is then well south of the Equator, and observers in southern latitudes will have the best chance of seeing this elusive planet.

Venus rises shortly before the Sun in northern latitudes, but it may still be seen in the south-east before sunrise in the Southern Hemisphere. Magnitude –3.3 to –3.4.

Mars is still an evening star, moving direct in Aquarius, but it sets shortly after the Sun and is not likely to be seen.

Jupiter rises in the late evening in the east in Virgo. The planet is now moving retrograde and on 19 February it passes about a degree south of Saturn for the second time, at a position a little west of that of the first conjunction in January. Jupiter continues to grow brighter as it approaches opposition (magnitude –1.8 to –1.9).

Saturn is also moving retrograde in Virgo, quite close to Jupiter. A feature of this year's night sky is the proximity of these two bright planets for the greater part of the year. The conjunction with Jupiter on 19 February is mentioned above. The magnitude of Saturn is +0.9 to +0.8, and both planets are bright enough to be easily recognized in a part of the sky containing few bright stars.

An annular eclipse of the Sun on 4 February will be visible as a partial eclipse in New Zealand and eastern Australia. For details see page 124.

JUPITER AND SATURN

If we regard the periods of Jupiter and Saturn as 12 and 30 years approximately, we see that Jupiter covers about 30 degrees a year, while Saturn moves through 12 degrees. Thus, Jupiter gains 18 degrees a year on Saturn, and conjunction of the two planets can only occur at an interval of 20 years. If both planets travelled in circular orbits it can be shown that only one in six of these conjunctions could possibly be triple, and we should then expect to have a triple conjunction every 120 years. However, both of these giant planets have eccentric orbits, and both are subject to severe perturbations, so that this average is never realized. The last three triple conjunctions of Jupiter and Saturn occurred in 1452, 1683 and 1940, and the intervals here are more than double the 120 years. In all of these cases, the time between the two oppositions was of the order of a day or less.

One triple conjunction of Jupiter and Saturn has received a great deal of attention. This is the conjunction of BC7 which, it was suggested, could be the explanation for the Star of Bethlehem. This old idea had been rejected in many quarters because the two planets were well separated in latitude and were, in any case, familiar objects to the Magi. In recent times the subject has been revived, but now the *astrological* significance of the event has been emphasized. This seems a more reasonable suggestion, though it does not explain all the details of the story. Certainly the rarity of this triple conjunction (which the Magi could never have witnessed before) would give added significance to the event.

BORRELLY'S COMET

This is one of the best-known of all periodical comets. It was first seen in 1905, and this year's return to perihelion (February 20) will be its tenth appearance. It is in the southern sky, but the magnitude may be not much below 15, so that it is within the photographic range of well-equipped amateurs. The period is 6.8 years.

GAMMA LEONIS

This is the second brightest star in Leo, and is contained in the famous Sickle. The magnitude is 1.99, so that it is half a magnitude fainter than Regulus. It is interesting to compare it with Denebola (Beta Leonis), whose present magnitude is 2.14. The difference between Beta and Gamma is almost inappreciable with the naked eye, but old reports give Beta as being of the first magnitude, in which case it has faded over the past two thousand years—though it is not a star of the type which would be expected to behave in this way (the spectral type is A3) and it would be unwise to trust the old records too far. Beta has, however, been suspected of variability, and it is always worth making a check.

Gamma Leonis, or Algieba, has a spectral type of K0, and is markedly orange. The distance from us is 90 light-years, and the absolute magnitude is +0.1, so that it is considerably more luminous than the Sun. However, its main interest lies in the fact that it is a fine binary. The G5-type companion is of magnitude 3.8, and the separation is 4.3 seconds of arc, so that the pair may be well seen with a small telescope. The orbital period is 407 years, and the separation is increasing slowly. This is certainly one of the best doubles in the sky from the viewpoint of the amateur using a small telescope.

Vesta, Minor Planet No. 4, comes to opposition this month not far from Gamma Leonis, but the magnitude is below 6, so that at this opposition Vesta is not likely to be seen with the naked eye.

THE SOUTH CELESTIAL POLE

Northern Hemisphere observers have an excellent pole star, Alpha Ursæ Minoris (Polaris). The Southern Hemisphere has only the insignificant Sigma Octantis, and the whole area is very barren. The best way to locate the South Pole is to look midway between Achernar, the leader of Eridanus, and the Southern Cross. During February evenings in the Southern Hemisphere both Achernar and the Cross are fairly high up, Achernar descending and the Cross rising, so that the polar region may be located without much difficulty.

March

New Moon: 6 March   *Full Moon:* 20 March

Summer Time in Great Britain and Northern Ireland commences on 22 March.

Equinox: 20 March

Mercury is at greatest elongation west (28 degrees) on 16 March and is then a morning star, but is not well placed for northern observers, the ecliptic being very low in the east at this time. In southern latitudes Mercury will be seen in the east, well above the horizon before sunrise. The planet is brightest at the end of the month, when it should be possible to see it setting in a dark sky.

Venus is approaching conjunction with the Sun and is not visible in northern latitudes. In the Southern Hemisphere it rises shortly before the Sun.

Mars is also nearing conjunction and will not be visible during the month.

Jupiter is moving retrograde in Virgo, and is at opposition on 26 March near the fourth magnitude star Eta Virginis. The planet is approaching aphelion, and at opposition is 414 million miles (666 million km) from the Earth. Although it reaches magnitude −2.0 it can be half a magnitude brighter than this at a September opposition when it is at perihelion.

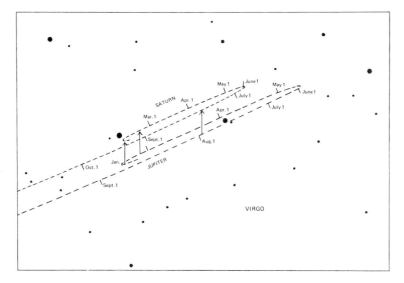

Saturn is at opposition on 27 March, less than 24 hours after the opposition of Jupiter. It is this small interval that causes the loops on the apparent paths of the planets to overlap, and gives rise to the three conjunctions. The diagram above shows the loops on a larger scale than that given in the January notes, and shows stars down to magnitude 6.5. Saturn is in Virgo to the east of Jupiter, and at opposition will be 799 million miles (1,280 million km) from the Earth. Its magnitude is now +0.6, but it can be much brighter than this when the rings are fully open. At present, the rings, which contribute much of the light of this planet, are only slowly opening again after being presented edge-on to the Earth and the Sun in 1980.

HYDRA, THE WATERSNAKE

This is a good month to look for Hydra, which is the largest constellation in the sky; it covers an area of 1,303 square degrees, slightly more than its nearest rival, Virgo (1,294 square degrees).

It is an immensely long constellation, extending from near Cancer almost as far as Libra. Most of it lies in the Southern

Hemisphere, though the 'head' is slightly north of the Equator; the brightest star in the head, Zeta Hydræ, has a declination of +6 degrees.

Despite its great size, Hydra is somewhat barren. There are only nine stars above the fourth magnitude, and only one above the second. This is Alphard (Alpha Hydræ), magnitude 1.98. It is not hard to identify, because a line drawn from Castor through Pollux and prolonged will pass not far from it; moreover, Alphard is very much 'on its own', and is often called the Solitary One. It has a K2-type spectrum, and the orange colour is noticeable even with the naked eye; binoculars bring it out well. The distance from us is 94 light-years, and the luminosity is 115 times that of the Sun.

It is interesting to recall that during his return to England from the Cape of Good Hope, in the 1830s, Sir John Herschel made observations of Alphard, and regarded it as definitely variable over a limited range. This has not been confirmed, and it is not now included in variable star lists, but fluctuations over a small range would not be surprising. To the naked-eye observer, the problem is one of finding suitable comparison stars at comparable altitude.

One interesting star which is unquestionably variable is R Hydræ, not far from the third-magnitude Gamma. It was actually the fourth variable star to be discovered, by Maraldi in 1704; at that time only three variables had been found—Mira, Algol, and Chi Cygni. R Hydræ is a typical Mira star, with an M-type spectrum. The period is 386 days, and the magnitude ranges between 4 and 10, so that at maximum it is an easy naked-eye object.

There are three Messier objects in Hydra; 48 (an open cluster), 68 (a globular) and 83 (a galaxy). Of these, M.48 is just visible with the naked eye, but it is not particularly rich, and may be difficult to locate. The position is R.A. 8h 11m, declination –05 degrees 38. The only easy double star of any note is Epsilon, in the 'head'; the magnitudes of the components are 3.5 and 6.9, and the separation is 2".9. The brighter component is itself a very close binary, with an orbital period of 15 years.

CELESTIAL PHOTOGRAPHY

It is a regrettable fact that photography through a telescope is almost impossible without an equatorially-mounted, clock-driven instrument. However, an ordinary camera may be used to take impressive star-trail pictures, and there is an excellent opportunity this month, because Jupiter and Saturn are close together in Virgo. It is interesting to take 'trail' pictures of the two from night to night, and see how the relative positions change. A useful reference star is Spica, which also is close by, and is slightly fainter than Saturn.

April

New Moon: 4 April   *Full Moon:* 19 April

Mercury is not likely to be seen in the Northern Hemisphere, but is still well displayed in southern latitudes, where it can be seen in the east before sunrise during the first half of April. Mercury is at its brightest towards the end of this period, but moves in to superior conjunction on 27 April.

Venus is in superior conjunction on 7 April, and will not be visible during the month.

Mars is in conjunction with the Sun on 2 April. By the end of the month the planet will be in Aries as a morning star, but it rises only a few minutes before sunrise and is not likely to be seen.

Jupiter is an evening star in Virgo and is visible for most of the night. The planet is still moving retrograde, and at the beginning of April it passes about a quarter of a degree north of the fourth magnitude star Eta Virginis. (See diagram in March notes.) Magnitude –2.0 to –1.9.

Saturn is also an evening star in Virgo, and, like Jupiter, its retrograde motion carries it past Eta Virginis at the end of April. Although the planet is fading as its distance increases (magnitude +0.6 to +0.8), and it is not as bright as Jupiter, it is much brighter than any of the neighbouring stars.

Pluto is at opposition on 13 April on the borders of Virgo and Boötes. The planet is not visible to the naked eye (magnitude

+14), but as a matter of interest, its position is shown on the diagram in the January notes. At opposition Pluto is 2,707 million miles (4,354 million km) from the Earth and 2,795 million miles (4,496 million km) from the Sun. If these figures are compared with those of Neptune in the June notes, it will be seen that Pluto's eccentric orbit has carried it nearer to the Sun than Neptune, although it is well north of the plane of the ecliptic, and there is never any chance of a close approach of the two planets.

THE NATURE OF PLUTO

Pluto was discovered fifty years ago by Clyde Tombaugh, at the Lowell Observatory in Arizona. The discovery was actually made in February, but the announcement was delayed until March to make sure that there could be no possibility of a mistake having been made. The success was not mere luck; Tombaugh had been carrying out a systematic search, using the 13-inch refractor which had been obtained specially for the task, and checking the plates upon a blink-comparator.

Pluto was found not far from the position given by Lowell. Yet from the start it has presented astronomers with puzzle after puzzle, and its nature is still uncertain. The fiftieth anniversary of the discovery, in 1980, was marked by a special Pluto conference at the New Mexico State University in Las Cruces, where Tombaugh is now Emeritus Professor. It was a great occasion; Tombaugh was presented with the University's highest award, the Regent's Medal, and was also honoured by having Asteroid No. 1604 named after him. This was one of the numerous asteroids which he had recorded during the search.

The main problem about Pluto is its small size and mass. According to recent estimates the mass is a mere 0.002 that of the Earth, and the diameter is less than that of the Moon. The overall density cannot exceed 1.3 times that of water, and the most probable composition is 21 per cent rock, 5 per cent methane and 74 per cent water ice. The surface seems to be coated with a layer of methane ice. Otherwise, all we really know is that the rotation period is likely to be 6.3 days.

What has really set astronomers thinking is the reported

discovery of a satellite. This was due to James Christy and his colleagues in 1977, following an examination of pictures of Pluto taken at the U.S Naval Observatory to check on the planet's motion; the aim was to see whether Pluto could be due to occult a star, which would enable its diameter to be determined by Gordon Taylor's method. The images were asymmetrical, and Christy attributed this to the existence of a satellite with a diameter one-third that of Pluto itself, moving in a period equal to that of Pluto's rotation. Not everyone is satisfied with this explanation; Dr Brian Marsden, at the conference, was somewhat sceptical, and regretted that the hypothetical body had already been given a name (Charon). The only real proof (or disproof) may come when the Large Space Telescope is launched in the mid-1980s; this should be able to photograph Pluto clearly enough to show whether or not Charon is a separate body. If confirmed, it will mean that the Pluto-Charon sytem must be regarded as a double planet.

Yet there are other possibilities. Pluto is much too small to be given planetary status; it is certainly smaller than Triton, the senior satellite of Neptune, and Marsden went so far as to suggest that it would be better classed as an exceptional asteroid. It could well be of the same nature as Chiron, which moves mainly between the orbits of Saturn and Uranus. There could in fact be many more similar bodies awaiting discovery.

One thing is certain: Pluto is not Lowell's Planet X. It could not possibly have caused measurable perturbations in the movements of giant worlds such as Uranus and Neptune. Either the discovery was sheer chance, or else there is another planet in those remote regions. Over his years of searching Tombaugh examined a grand total of 19 million star images, and he could hardly have missed another planet no fainter than Pluto; but the last word has by no means been said, and even today, after half a century, we do not really know the kind of body that Pluto is.

KOHOUTEK'S PERIODICAL COMET

In 1975, Lubos Kohoutek, at Hamburg, discovered a faint comet which is periodical—and not to be confused with

Kohoutek's Comet of 1973, which was expected to become brilliant, but signally failed to do so! The periodical comet is due at perihelion on April 17, but as it is only the 17th magnitude it is well beyond the range of amateurs. The period is 6.2 years.

May

New Moon: 4 May   *Full Moon:* 19 May

Mercury is at greatest eastern elongation (23 degrees) on 27 May, and is then an evening star, well placed for observation in northern latitudes, where it will be seen to the north of west after sunset. The diagram shows the changes in altitude and azimuth (true bearing from the north through east, south and west) of Mercury on successive evenings when the Sun is six degrees below the horizon in latitude 52 degrees north (about 40 minutes after sunset at this time of year). The changes in brightness are roughly indicated by the size of the circles, and it

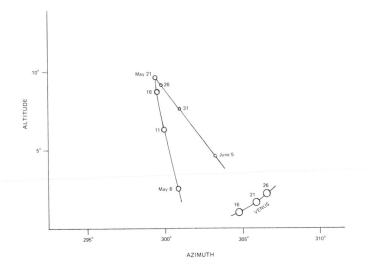

will be seen that Mercury is brightest *before* it reaches greatest elongation. Conditions are less favourable in southern latitudes, but it should be possible to see the planet in the northwest after sunset at the end of the month.

Do not be confused by the presence of Venus in the same part of the sky—Venus is much brighter and sets earlier than Mercury.

Venus is now an evening star and is well north of the Equator. For northern observers the planet will be seen to the north of west, setting about an hour after the Sun at the end of May. Magnitude –3.4. In the Southern Hemisphere, the planet is not likely to be seen until the end of the month.

Mars is a morning star in Aries, but it is very slow to move outwards from the Sun, and is not likely to be seen in the Northern Hemisphere. It is better placed in southern latitudes, and by the end of May it will have moved into Taurus, and will rise about an hour before the Sun. Magnitude +1.5 to +1.6.

Jupiter is an evening star, setting in the west in the early morning hours. The planet is moving retrograde in Virgo, but reaches its second stationary point on 28 May at the western end of the constellation. As will be seen from the diagrams (January and March notes), the retrograde path of Jupiter is longer than that of Saturn, but it covers the greater distance in less time—retrograding through ten degrees in 123 days as compared with Saturn covering seven degrees in 138 days. Magnitude of Jupiter at the end of May,–1.8.

Saturn sets shortly after Jupiter and will be found about three degrees to the east of the brighter planet. Saturn is moving slowly retrograde in Virgo as it approaches its second stationary point. Magnitude +0.9 to +1.0.

Uranus is at opposition on 19 May in Libra. The planet is just visible to the naked eye (magnitude +5.8) and in a small telescope it appears as a greenish disk. The position of Uranus is shown on the diagram in the June notes. At opposition, Uranus is 1,656 million miles (2,664 million km) from the Earth.

THE NAMING OF URANUS

Uranus, the first planet to be discovered since ancient times, was detected by William Herschel on 13 March 1781. Herschel was engaged in a 'review of the heavens', and was not looking for planets; he did not even recognize what he had found, and believed it to be a comet. The first recognition of it as a planet seems to have been due in the following May (two centuries ago this month) to two mathematicians. One was J. de Saron, a French aristocrat who was a colleague of Charles Messier; de Saron was later guillotined during the Revolution. The other was the Finnish mathematician Anders Lexell, who gave an orbital period of between 82 and 83 years and a distance from the Sun of 19 astronomical units, which is not far from the truth.

Once the planet had been identified, the question of a name came to the fore. Johann Elert Bode was one of the first in the field, and it was he who suggested Uranus, following the mythological tradition; Uranus was the father of Saturn. About the same time J. Bernouilli proposed 'Hypercronius'. In 1782, Herschel proposed 'the Georgian Planet', in honour of his royal patron, King George III of England and Hanover. Not surprisingly, this was received with a marked lack of enthusiasm, and many astronomers referred to the planet simply as 'Herschel'.

These various names were used for a surprisingly long time. As recently as 1849, well after the discovery of Neptune, the *Nautical Almanac* was still keeping to 'the Georgian'. In the end, mythology prevailed. John Couch Adams proposed changing over to Bode's original suggestion of Uranus; the *Nautical Almanac* did so, and the name became universally accepted.

At an early stage it was clear that the new planet was a giant. Lexell, in 1781, stated that it was larger than any other planet apart from Jupiter and Saturn. In 1788, Herschel himself measured the diameter as 34,217 miles or 55,067 kilometres, and it is now believed that this may be very close to the truth; very recent measurements make Uranus larger than had been widely believed in our own century, and it is slightly but perceptibly larger than Neptune, though less massive and less dense.

THE SOUTHERN CROSS

To observers in South Africa, Australia and New Zealand, the Southern Cross—Crux Australis—is near the zenith during May evenings, and it is as familiar to Southern Hemisphere dwellers as the Great Bear is to Europeans and North Americans.

It may come as a surprise to learn that Crux is the smallest constellation in the sky, and was not even recognized as an independent group until 1679; previously it had been included in Centaurus. Neither is it really like a cross. It is more like a kite. There are four main stars; Alpha, Beta, Gamma and Delta, while a fifth (Epsilon) rather spoils the symmetry. Alpha, a spendid double, is called Acrux, clearly a 'made-up' name; Beta has the unofficial name of Mimosa.

It is interesting to compare the apparent and absolute magnitudes of the four leaders (absolute magnitude, it may be recalled, is the apparent magnitude that a star would have if seen from a standard distance of 10 parsecs, or 32.6 light-years). The figures are as follows:

Star	Apparent magnitude	Absolute magnitude
Alpha	1.4, 1.9	–3.9, –3.4
Beta	1.3	–4.6
Gamma	1.7	–2.5
Delta	2.8	–3.4

Alpha appears the brightest member, with a combined magnitude of 0.8; Delta seems much the faintest, but only because it is further away (570 light-years). Gamma Crucis is a red giant of spectral type M3, while the others are hot and white. The difference between Gamma and its companions is noticeable at a glance.

Crux is exceptionally rich in interesting objects; note particularly the "Jewel Box" round Kappa Crucis, a glorious open cluster, and the dark nebula known as the Coal Sack. European and United States astronomers never cease to regret that Crux is so far south in the sky!

June

New Moon: 2 June   *Full Moon:* 17 June

Solstice: 21 June

Mercury may still be seen as an evening star in southern latitudes in the first few days of June, but the planet is not very bright. Mercury passes less than two degrees south of Venus on 9 June and is in inferior conjunction on 22 June.

Venus is an evening star, setting to the north of west more than an hour after the Sun. Magnitude –3.3.

Mars is a morning star in Taurus and at the beginning of June it will be about three degrees south of the Pleiades. The planet passes six degrees north of Aldebaran on 19 June. By the end of the month Mars rises more than an hour before the Sun, but the planet is not very bright (magnitude +1.6 to +1.7). This may be judged by comparing its brightness with that of Aldebaran (magnitude +1.1).

Jupiter is an evening star, setting in the west about midnight. It is now moving direct in Virgo, and is gradually approaching conjunction with Saturn. Magnitude –1.7 to –1.6.

Saturn is also an evening star, about two degrees to the east of Jupiter. The planet reaches its second stationary point on 6 June, and after this it begins to move very slowly direct once more. Magnitude +1.1 to +1.2.

Neptune is at opposition on 14 June in Ophiuchus. The planet

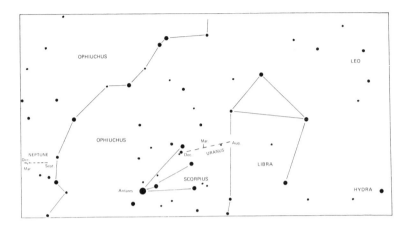

is not visible to the naked eye (magnitude +7.7), but its position at this time is shown in the diagram above. The distance of Neptune from the Earth at opposition is 2,723 million miles (4,379 million km) and it is 2,816 million miles (4,529 million km) from the Sun. These figures change very little from year to year since the orbit of Neptune is almost circular, but comparison with the figures given for Pluto in the April notes will show that Pluto is now much nearer to the Sun than Neptune. It must not be thought that Pluto is 'inside' the orbit of Neptune, because it is actually some 800 million miles north of the plane of Neptune's orbit.

THE RING OF NEPTUNE

Saturn's rings have been known for over three centuries; they were first seen by Galileo in 1610 (even though he did not recognize them for what they were), and were identified as rings by Christiaan Huygens less than fifty years later. The ring-system of Uranus was detected in 1977, by the occultation method (it is surprising how often we come back to Gordon Taylor's technique for measuring the apparent diameters of faint Solar System bodies!) and the Jovian ring was found from the Voyagers of 1979.

Of course, these systems are not alike. Saturn's rings are in a class of their own, and it is generally admitted that Saturn is the loveliest object in the entire sky. There has been visual confirmation of the rings of Uranus, but not by direct telescopic viewing; it is not likely that the thin Jovian ring can be seen from Earth—and the natures of these two systems are quite unlike those of Saturn. Nevertheless, rings exist. What, then, about Neptune?

Neptune, at opposition this month, was discovered in 1846 as a result of calculations made by the French mathematician Urbain Le Verrier. The identification was made at Berlin by Johann Galle and Heinrich D'Arrest. At Cambridge, James Challis had been searching on the basis of similar calculations made by John Couch Adams, and had failed to locate Neptune only because of his laborious search procedure and, it must be admitted, because of his own incompetence and lack of enthusiasm. Soon afterwards, the English amateur William Lassell discovered Triton, Neptune's senior satellite. There were also reports of a Neptunian ring.

Similar suspicions had attached to Uranus long before, and William Herschel himself had believed—for a while, at least—that rings existed. We now know that he was deceived by his optical arrangement; he was using a 'front-view' reflector, tilting the main mirror and dispensing with the flat. This may sound logical, but in fact it presents all manner of problems, and it is unlikely that any Herschelian reflectors are in use today. The Uranian rings discovered in 1977 are not those of Herschel, which did not exist.

The Neptunian ring proved to be equally ghostlike. Challis thought he had seen it; so did Lassell, but Lassell then became dubious. Since then, photographs have established that it, too, is non-existent.

Yet can we be sure that Neptune is ringless? Probably not. If a ring-system exists, it must be of the same type as those of

Jupiter and Uranus, but it may be there. The only chance of detecting it will be by space-probe, and we must await the fly-by of Voyager 2 in the late 1980s. If all goes well, we will then find out whether or not Neptune, like the other giants, is a ringed world.

A GREEN NAKED-EYE STAR?

Libra, the Scales or Balance, is one of the most obscure of The Zodiacal constellations (it was originally known as Chelæ Scorpionis, the Scorpion's Claws). There are only two stars above the third magnitude, and three more above the fourth; neither is there any distinctive pattern.

The brightest star is Beta Libræ, which has the somewhat tongue-twisting proper name of Zubenelchemale. It is of magnitude 2.6; the absolute magnitude is –0.6, the distance from us 140 light-years, and the spectral type B8. It is of mild interest because it is often said to be the only single star with a decidedly greenish hue.

This may or may not be so. Most observers (including the Editor!) will describe it as white. However, the reports have been fairly persistent, and it is worth taking a look at Beta Libræ to see whether any greenish tinge can be detected.

July

New Moon: 1 and 31 July   *Full Moon:* 17 July

Earth is at aphelion (farthest from the Sun) on 3 July, when its distance will be 94.5 million miles (152.1 million km).

Mercury is at greatest elongation west of the Sun (21 degrees) on 14 July. It is then a morning star and in northern latitudes may be seen a few degrees above the horizon to the north of east before sunrise. Conditions are rather better in the Southern Hemisphere, and Mercury will be well above the horizon in the middle of July, although much brighter towards the end of the month.

Venus is an evening star and in the Northern Hemisphere it sets about an hour after the Sun. The planet is slowly increasing its elongation from the Sun, but at the same time it is moving south, so that conditions are improving in southern latitudes, but remain much the same in the north. As a result, Venus continues to set about an hour after the Sun for the next three months in the Northern Hemisphere, but at 35 degrees south the planet will set in a dark sky about two hours after the Sun at the end of July. Magnitude −3.4.

Mars is a morning star moving direct in Taurus, and at the end of July it passes into Gemini. It will then rise more than two hours before the Sun (northern observers) but rather less in the Southern Hemisphere. Magnitude +1.7 to +1.8.

Jupiter is still an evening star in Virgo, but now sets in mid-evening, and at the beginning of July it will be seen to be little more than a degree west of Saturn. Both planets are now moving direct, but Jupiter will overtake Saturn on 30 July, passing rather more than a degree south of Saturn at this third conjunction. Magnitude at this conjunction –1.4. Jupiter will be at aphelion on 28 July.

Saturn is in Virgo in the east of Jupiter, and is overtaken by that planet on 30 July. Magnitude +1.2. Both planets are close to the Equator and set in the west in mid-evening at all latitudes.

A partial eclipse of the Moon on 17 July will be visible in North and South America. Further details on page 124.

A total eclipse of the Sun on 31 July will be visible in the U.S.S.R. and the North Pacific Ocean. Visible as a partial eclipse in most of Asia and in the extreme north-west of North America. See page 125.

THE ORBITS OF THE PLANETS

The Earth reaches its aphelion on July 3. It had passed its perihelion on January 2. The distance has ranged between 91.4 million miles and 94.5 million miles (147 million kilometres and 152 million kilometres). The orbit is, then, appreciably eccentric; but it is worth making a few comments, because some popular misconceptions still persist.

First, it is quite wrong to suppose that the orbit is egg-shaped, as sometimes depicted in books. The total range is only just over 3 million miles, whereas the mean distance between the Earth and the Sun is 93 million miles; 3 parts in 93 is not very much! In fact, drawn to a small scale, the orbit will appear virtually circular.

Other planetary orbits have different eccentricities. Note, also, that the values are not quite constant, because of mutual perturbations. The values of epoch 1980 are as follows:

Mercury	0.206	Saturn	0.056
Venus	0.007	Uranus	0.047
Earth	0.017	Neptune	0.009
Mars	0.093	Pluto	0.248
Jupiter	0.048		

Pluto stands out at once; the distance from the Sun ranges between 7,375 million kilometres and 4,425 million kilometres, but, as we have noted, there are grounds for demoting Pluto from its planetary status. Of the rest, Mercury has the most eccentric orbit, Venus and Neptune the least.

So far as larger planetary satellites are concerned, the eccentricities are usually very low, and the orbits are very nearly circular. With some of the 'asteroidal' satellites the figures are greater; the record is held by Nereid, the dwarf attendant of Neptune, where the eccentricity is 0.749. This is more like a cometary orbit than a planetary one, and has led to (or, at least, supported) suggestions that something very strange happened long ago in the far reaches of the Solar System—so that Pluto was wrenched free from Neptune, Triton thrown into a retrograde orbit and Nereid into a cometary one. Of course, these suggestions are very tentative, and many astronomers are highly sceptical about them, but it is certainly strange that of large satellites only Triton should revolve in a direction opposite to that of the spin of its primary.

It is often asked: Why are all orbits elliptical rather than circular? The answer is quite straightforward. The eccentricity of an orbit is defined by the ratio between the major axis of the orbit, and the distance between the two foci. The closer together the foci, the smaller the eccentricity. When the foci coincide, the eccentricity is zero. In other words, a circle is merely an ellipse with eccentricity 0.0, so that it is a special case!

URSA MINOR

Ursa Minor, the Little Bear, is celebrated as being the constellation to include the north celestial pole. The declination of Polaris (Alpha Ursæ Minoris) is $+89°$ 10′, so that it is within a degree of the actual pole. Its magnitude is 1.99; it is actually a Cepheid variable, but the fluctuations are too slight to be detected with the naked eye. Polaris is 680 light-years away, and has an absolute magnitude of –4.6, so that if it were closer to us it would be a really brilliant object. The spectral type is F8, so that the surface is slightly hotter than that of the Sun.

The other bright star in the constellation is Beta, or Kocab (magnitude 2.04); here the spectral type is K5, and the colour is clearly orange, as binoculars will show well. It and its third-magnitude neighbour, Gamma (Pherkad Major) are sometimes known as the Guardians of the Pole. Actually there are not many interesting telescopic objects in Ursa Minor, but the shape is distinctive enough on a clear night; it is not unlike a faint and distorted version of Ursa Major.

August

Full Moon: 15 August   *New Moon:* 29 August

Mercury is at superior conjunction on 10 August, and will not be seen until the end of the month, when it begins to appear as an evening star. Although quite bright, it will be very low in the western sky in northern latitudes, but will be well displayed in the Southern Hemisphere, shortly after sunset.

Venus is an evening star and continues to set about an hour after sunset in northern latitudes. Observers south of the Equator will be more fortunate, the planet setting more than two hours after the Sun in a dark sky. Magnitude –3.4 to –3.5. Venus will pass two degrees south of Saturn on 25 August and less than a degree south of Jupiter on 28 August.

Mars is a morning star in Gemini and rises about two hours after midnight at 52 degrees north. It is now at its highest northern declination, and this reduces its visibility in the Southern Hemisphere, where it does not rise until dawn. The planet is not easy to identify as it is now at magnitude +1.8. Mars passes six degrees south of Pollux on 23 August, and at the end of the month passes into Cancer.

Jupiter now sets in the early evening as twilight ends, but is rather more impressive in the Southern Hemisphere. Jupiter will be seen with Saturn south of the crescent moon on 4 August, and by the end of the month the two planets will be joined by Venus. The two conjunctions on 25 and 28 August are mentioned above. Magnitude of Jupiter –1.3.

Saturn remains near Jupiter and also sets as twilight ends, but rather later in southern latitudes. The rapid change in the grouping of these two planets with Venus at the end of the month should be well worth observing. Magnitude of Saturn +1.2.

ZENITHAL HOUR RATES OF METEOR SHOWERS

August sees the return of the Perseid meteors, which make up the most consistent shower of the year. The maximum is due on or about August 12, when the ZHR or Zenithal Hourly Rate may exceed 65. Unfortunately the Moon will then have started to become obtrusive, but it will be absent from the •evening sky during the earlier stages of the shower, and observers will have favourable conditions—weather permitting!

The ZHR is the number of naked-eye meteors per hour which an observer may be expected to see, assuming perfect conditions and with the radiant at his zenith. In practice, of course, these conditions are never attained, but the ZHR is a good guide even though the actual rate of meteors recorded will always be rather less. The approximate ZHRs for some of the annual streams are as follows:

Shower	Date of Maximum	ZHR
Quadrantids	January 4	variable – up to 250
Corona Australids	March 16	5
Lyrids	April 21	12
Aquarids	May 5	20
Ophiuchids	June 20	6
Delta Aquarids	July 25	35
Piscis Australids	July 31	8
Capricornids	August 2	8
Perseids	August 12	68
Orionids	October 20	30
Taurids	November 8	12
Leonids	November 17	variable
Geminids	December 14	58
Ursids	December 22	5

The Quadrantids have a short, sharp maximum, and are described in detail by John Mason elsewhere in this *Yearbook*. The Leonids are the most erratic of the annual showers. They provided spectacular displays in 1799, 1833 and 1866, and again in 1966, but in most years they are feeble, with a ZHR which does not exceed 10. Both the Taurids and the Geminids are rich in fireballs (that is to say, very bright meteors) but it is the Perseids which may be relied upon to give the most regular display of the year.

THE NORTHERN CROSS

One of the most splendid constellations in the entire sky is Cygnus, the Swan. It is often nicknamed the Northern Cross, and it really is shaped like an X, so that it is far more cruciform then the Southern Cross. It is also much larger, but admittedly its stars are not so brilliant. One, Deneb, is of the first magnitude; there are three more above the third, and a total of eleven above magnitude 4.

Deneb is a true celestial searchlight. It is over 1,500 light-years from us, and its luminosity may be as high as 60,000 times that of the Sun; the spectral type is A2. The symmetry of the X is rather spoiled by the fact that one member, Beta Cygni or Albireo, is fainter than the rest (magnitude 3.1) and is further away from Gamma, the centre of the cross; but to atone for this, Albireo is a superb double star—without much doubt the finest in the sky. The primary is golden-yellow, the companion blue; the separation is nearly 35 seconds of arc, and since the companion is of magnitude 5.4 the pair may be separated even with good binoculars. It has recently been found that there is a third member of the system, but only the two famous stars are visible telescopically.

Cygnus contains many interesting objects. There is the unstable P Cygni, a 'pseudo-nova', which is now of about the fifth magnitude; there is the Mira variable Chi Cygni, which may exceed magnitude 4 at maximum; there is also SS Cygni, the brightest of the U Geminorum variables or dwarf novæ, which can be followed with very modest telescopes. Moreover, Cygnus lies in a rich part of the Milky Way, and there are innumerable glorious starfields. Various novæ have appeared in the constellation; the last, V.1500 Cygni of 1975, reached magnitude 1.8, but it faded very quickly, and has now become a difficult telescopic object.

September

Full Moon: 14 September   *New Moon:* 28 September

Equinox: 23 September

Mercury is at greatest eastern elongation (26 degrees) on 23 September, and is then an evening star. For observers in the Northern Hemisphere the planet sets shortly after the Sun, but it will be well displayed for those in southern latitudes. Here the planet will be seen well above the horizon in the west about half an hour after sunset, and the sky will be dark by the time Mercury sets.

Venus is an evening star and is now south of the Equator, so that it is not well placed for northern observers who will only see the planet for about an hour after sunset. In the Southern Hemisphere Venus sets in a dark sky about three hours after the Sun, and continues to grow brighter as it moves out from the Sun. Magnitude –3.5 to –3.6.

Mars is a morning star in Cancer and in northern latitudes it rises about two hours after midnight in a dark sky. Southern observers will not see the planet until it rises at dawn. The diagram shows the path of Mars for the rest of the year. From now on the planet grows rapidly brighter as its distance decreases. Magnitude +1.8.

Jupiter is approaching conjunction and sets in the twilight sky shortly after sunset. The planet is now south of the Equator, and for the next few years will be well placed for observers in the Southern Hemisphere.

Saturn is still near Jupiter but will be difficult to see in the bright twilight (magnitude +1.2 to +1.1). The remarks given above for Jupiter apply to Saturn also.

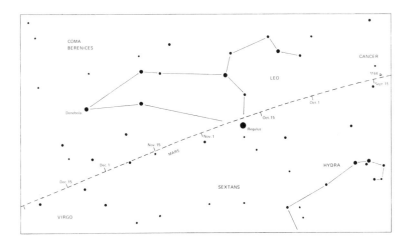

THE SOUTHERN BIRDS

Far in the Southern Hemisphere of the sky, and therefore permanently below the horizon of Britain or the United States, lie the four Southern Birds—Grus (the Crane), Pavo (the Peacock), Tucana (the Toucan) and Phœnix (the Phœnix). Apart from Grus they are not very distinctive, but all have their own special points of interest, and the map given here will help in identifying them; admittedly this is a rather confusing area of the sky.

Grus is, however, fairly easy to recognize on two counts. First, it has a pair of brightish stars; Alpha or Alnair (magnitude 1.8) and Beta or Al Dhanab (2.1). Even a casual glance shows that they are not alike; Alnair is white, Al Dhanab a lovely warm orange. The spectra are respectively B5 and M3. It is instructive to look at them alternately, using binoculars; the contrast is very marked indeed. Al Dhanab is a member of a line of faint stars, of which two (Delta and Mu Gruis) give the

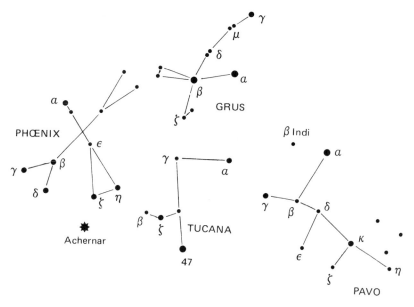

impression of being wide doubles, though in fact the components are not physically associated, and lie at different distances from us.

Phœnix has one brightish star, Alpha or Ankaa (magnitude 2.4). Zeta Phoenicis is an eclipsing binary; the range is from magnitude 3.6 to 4.1, so that the variations are easy to follow with the naked eye. The period is short, only 1.7 days. A good comparison star is Delta, magnitude 4.0.

Pavo also has one bright star, Alpha (magnitude 1.9) which is rather separated from the rest of the constellation. The most interesting object is the short-period variable Kappa Pavonis, which is of the W Virginis type (allied to the more regular Cepheids). Kappa ranges between magnitudes 4.0 and 5.5 in a period of 9.1 days. It is highly luminous, and lies at a distance of 650 light-years. Delta Pavonis, of magnitude 3.6, is a mere 19 light-years away, and is very similar to the Sun; recent doubts have been expressed as to the probability of other life-bearing planets in our part of the Galaxy (see the article by

James Oberg!); but as a centre of a solar system, Delta Pavonis is a far from unpromising candidate.

Finally, there is Tucana, the least bright of the Birds, but distinguished by the presence of the globular cluster 47 Tucanæ, inferior only to Omega Centauri. Even with binoculars it is a glorious object, and it is dimly visible with the naked eye as a misty patch. Note also Beta Tucanæ, which is a wide double; each component is again double. Alpha Tucanæ, the brightest star in the constellation, is of magnitude 2.7.

FOMALHAUT

For Northern Hemisphere observers, this is a good time to look for Fomalhaut in the Southern Fish, the southernmost of the first-magnitude stars visible from Britain or the northern United States. It may be found from the Square of Pegasus, and is bright enough to be conspicuous; its declination is $-29°44'$.

October

Full Moon: 13 October   *New Moon:* 27 October

Summer Time in Great Britain and Northern Ireland ends on 25 October.

Mercury is in inferior conjunction on 18 October and will not be seen until the end of the month, when it begins to appear as a morning star. Conditions are best in the Northern Hemisphere, and although the planet is not very bright, it should be seen low in the south-east before sunrise. See November notes.

Venus is an evening star and is growing brighter as it approaches eastern elongation (magnitude −3.7 to −3.9). At northern stations Venus will be visible for an hour or more after sunset, but in southern latitudes it makes a more impressive appearance, setting nearly three hours after sunset to the south of west.

Mars moves from Cancer into Leo at the beginning of the month and will pass rather more than a degree north of Regulus on 19 October. It then rises about an hour after midnight in the Northern Hemisphere, but will not appear until shortly before dawn at 35 degrees south. Magnitude +1.7 to +1.6.

Jupiter is in conjunction with the Sun on 14 October, but by the end of the month it begins to appear as a morning star in the east just before sunrise. This will be a difficult observation in the bright dawn sky of the Northern Hemisphere, but the earlier dawn of southern stations will prevent observers from seeing this planet for some time yet.

Saturn is in conjunction with the Sun on 6 October. After this it becomes a morning star, but is not very bright as yet (magnitude +1.0), and is not likely to be seen in the bright dawn sky.

'MISLAID' COMETS

This month, Longmore's periodical comet is expected to return. The date of perihelion is given as 21 October, but this must be regarded as somewhat uncertain. The comet was seen originally on 10 June 1975, and it was followed until 4 October, but only eight observations of it were made, so that the orbit is not at all well known. Moreover, the magnitude can hardly become much brighter than 16. Whether or not it will be recovered remains to be seen.

There are various comets which have been assumed to be periodical, but which have been seen at one apparition only. The causes are varied. In some cases (Lexell's Comet of 1770, in particular) planetary perturbations have been so great that all track of the comet has been lost. Other comets may have disintegrated. It may simply be that the predictions have not been sufficiently accurate. Of course, there are some comets which we may confidently expect to see again; thus, Kowal's Comet of 1977 has a period of 15 years, so that, all being well, it will return once more in 1992.

Neither is it ever safe to give up a comet as being lost. Brian Marsden, English-born but now resident in the United States, has a fine record of theoretical recoveries of comets, and some unexpected discoveries have been made; thus Denning's Comet of 1881 has a period of 9 years, but was not picked up again until the Japanese astronomer Fujikawa recovered it in 1978.

The following are some notable comets which have been seen at only one return:

Comet	Year seen	Period, years.
Helfenzrieder	1766	4.5
Barnard 1	1894	5.4
Brooks 1	1884	5.6
Lexell	1770	5.6
Pigott	1783	5.9

Spitaler	1880	6.4
Barnard 3	1892	6.6
Schorr	1918	6.7
Swift 2	1895	7.2
Denning 2	1894	7.4
Peters	1846	13.4
Perrine	1916	16.4
Pons-Gambart	1827	63.9
Swift-Tuttle	1862	119.6
Mellish	1917	145.3

How many of these will be recovered? Lexell's, which approached the Earth more closely than any other known comet and was visible with the naked eye, certainly will not; its new orbit leaves it too far from the Earth, and its period has been lengthened. On the other hand, there is no reason why Mellish's Comet should not be seen again in A.D. 2062. This, incidentally, is also the year that Halley's Comet will come back after its next appearance in 1986, but by then any present readers of this *Yearbook* will have become at least octogenarians!

BETA CETI

The second-magnitude star Beta Ceti (Diphda) is of some interest, because in the past it has been suspected of variability. The official magnitude is 2.0, but occasional reports of brightening have been made, and it is perhaps worth keeping a watch. Diphda, like Fomalhaut, may be found by using the Square of Pegasus as a guide. There can be little chance of confusing the two; as seen from northern latitudes Fomalhaut is lower down in the sky, and is considerably the brighter.

F. J. M. STRATTON

This is the centenary of the birth of the distinguished Cambridge astronomer Frederick John Marrian Stratton. He was born in Birmingham on 16 October 1881, but was associated with Cambridge for much of his life; he became Professor of Astrophysics there, and Director of the Solar Physics Observatory in 1928. During his long career Stratton made valuable contributions in studies of the Sun, variable stars and novæ.

November

Full Moon: 11 November   *New Moon:* 26 November

Mercury is at greatest western elongation (19 degrees) on 3 November, and northern observers will have a good opportunity of observing this planet as a morning star in the south-east before sunrise. The diagram shows the changes in altitude and azimuth of Mercury on successive mornings when the Sun is six degrees below the horizon; this is about 35 minutes before sunrise at 52 degrees north. The increase in brightness by the middle of November is quite marked, and is roughly indicated by the size of the circles. Mercury passes about a degree north

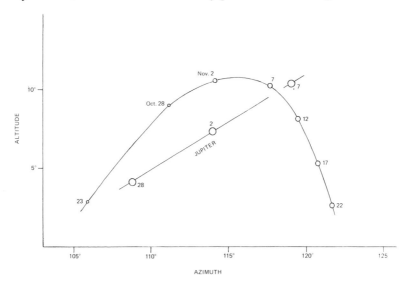

117

of Jupiter on 6 November, but there should be little risk of confusion, Jupiter being much brighter than Mercury and at a greater altitude after this date. In the Southern Hemisphere, Mercury will be very low in the south-east at sunrise.

Venus is at greatest eastern elongation (47 degrees) on 11 November and will be seen as an evening star for about two hours after sunset in northern latitudes. It is growing brighter (magnitude –4.0 to –4.2) and in a small telescope or binoculars it can be seen to resemble the Moon at First Quarter. In early November, Venus is nearly 27 degrees south of the Equator, so that it is well placed in the Southern Hemisphere, setting in a dark sky about three hours after sunset.

Mars is still a morning star moving direct in Leo, and rising about an hour after midnight at 52 degrees north. It is not so well placed for southern observers, and in these latitudes it will be visible for only an hour before dawn. Magnitude +1.6 to +1.4.

Jupiter now rises at dawn in northern latitudes and more than an hour earlier at the end of the month. In the Southern Hemisphere, sunrise and dawn are much earlier, and the planet may be lost in the bright sky. Jupiter has moved far ahead of Saturn and is now east of Spica, moving direct towards Libra. Magnitude –1.2 to –1.3.

Saturn is also a morning star, but is still west of Spica and therefore rises before Jupiter. The magnitude of Saturn is now +1.0 and it should be seen to rise in a dark sky at the end of November. The positions of both planets are shown in the diagram in the January notes.

COMET KWERNS-KWEE

This rather picturesquely named periodical comet was originally discovered in 1963 at Palomar, and was followed until the early part of 1965. It was again seen at the return of 1971. The period is 9 years, and perihelion is due on 30

November this year. Though the comet is extremely faint, its orbit is fairly well established, and there seems no reason why it should not be recovered on schedule.

THE LEONIDS

As has been noted earlier in this *Yearbook,* the Leonid meteor shower is very erratic; sometimes there are spectacular displays, while at other times, the shower is barely noticeable. Little can be expected in 1981, but if any major activity takes place it is likely to be on the early morning of 17 November. Meteor enthusiasts will certainly be on the alert, though more in hope than in anticipation.

MERCURY WITH THE NAKED EYE

Early November is a good time for Northern Hemisphere observers to look for the elusive planet Mercury, and 6 November is an ideal date, as Mercury will be very close to Jupiter in the sky. Actually, Mercury is brighter than most people expect. The maximum magnitude is –1.9, so that the planet is brighter than any star—even Sirius; but, of course, it is never seen against a dark background.

In general, Mercury is a naked-eye object on quite a number of occasions during a year, but it does need seeking. One thing which should emphatically *not* be done is to sweep around for it with binoculars when the Sun is above the horizon; it is only too easy to look at the Sun by mistake, with disastrous results. When the Sun has set, binoculars will naturally help; but put them away the instant that the first segment of the Sun appears above the horizon.

Telescopically, very little can be seen on Mercury even with powerful telescopes. Early maps were drawn by observers of the calibre of G. V. Schiaparelli and E. M. Antoniadi, but have proved to be very wide of the mark—which is no criticism of the observers! Our only detailed information has been gained from the Mariner 10 probe, which made three active passes of Mercury (March and September 1974, and March 1975) before going out of contact permanently. Almost half the surface has been mapped, and Mercury has proved to have features superficially similar to those of the Moon.

However, the owner of a modest telescope will at least be able to follow the changing phase, and this alone is a source of personal satisfaction. In the case of Venus, theoretical and observed phase are often in some disagreement; this is attributed to the dense atmosphere of Venus. The effect should be absent for Mercury, but some discrepancies have been recorded, and the problem is worth following up.

NOVEMBER CENTENARIES

Two centenaries are worth noting this month. The Italian astronomer Giovanni Antonio Amadeo Plana was born at Vogheri in Piedmont on 8 November 1781, and was educated in Paris; he became Professor of Astronomy in Turin, and carried out important studies of the Moon's motion, published in 1832. A lunar crater is named in his honour. He died on 20 January 1864.

Jean Alfred Gautier was born at Geneva, Switzerland, on 19 July 1793; he too was educated in Paris, and in 1819 he joined the staff of the Geneva Academy, becoming the first Director of the new Geneva Observatory. Unfortunately, eyesight trouble forced his retirement in 1837, but he continued his theoretical work, and was an independent discoverer of the identity of the solar and magnetic cycles (1852). Gautier died on 30 November 1881.

December

Full Moon: 11 December   *New Moon:* 26 December

Solstice: 21 December

Mercury is in superior conjunction on 10 December and will not be visible during the month.

Venus is at its greatest brilliancy on 16 December (magnitude –4.4). It is then a splendid object in the evening sky, setting more than two hours after the Sun. It is now moving in towards inferior conjunction in January 1982, and after this will be a morning star, reaching its greatest western elongation in April.

Mars moves into Virgo in early December, and by the end of the month it crosses the Equator on its way south. It will then rise at midnight in all latitudes. Mars grows rapidly brighter as its distance from the Earth decreases (magnitude +1.3 to +0.9) and it will reach opposition in March 1982.

Jupiter is a morning star moving direct in Virgo and approaching the borders of Libra. In the Northern Hemisphere the planet rises about an hour before dawn at the beginning of December, and more than an hour earlier at the end of the year. Southern observers are still at a disadvantage, the planet rising at dawn in early December. Jupiter grows a little brighter (magnitude –1.3 to –1.4) as it approaches opposition in April 1982.

Saturn is also a morning star, but rises earlier than Jupiter. The planet is moving direct and by the end of the year it is to be found north of Spica, but rather brighter than that star.

(Magnitudes Saturn +0.9, Spica +1.2). The next opposition of Saturn is in early April 1982.

VENUS AS A CHRISTMAS OBJECT

Venus was at its greatest elongation from the Sun last month, but it is at its most brilliant in mid-December. This may seem curious, but there is a good reason for it. Since Venus is an inferior planet, closer to the Sun than we are, it is in some ways awkward to observe. At its closest to the Earth it is 'new'; its dark side is turned toward us, and we cannot see the planet at all (except during the rare occasions of a transit across the Sun's disk; this last happened in 1882, and will not occur again until 2004—it is most improbable that there is anyone now living who has seen a transit of Venus!). When the phase is full, Venus is at superior conjunction on the far side of the Sun, and again is out of view. As the distance from Earth decreases, so does the phase, while the apparent diameter grows. Owing to the presence of the cloud-laden atmosphere, the increase in apparent diameter more than compensates for the decrease in phase, so that Venus is at its brightest when in the crescent stage. This is not true for the virtually atmosphereless Mercury, which is brightest when gibbous (that is to say, between half and full).

Venus is a deceptive world. Little can be seen on it apart from vague, cloudy and impermanent shadings; we now know that the surface is intolerably hot, with a temperature of the order of 900° F, and that the atmosphere is composed chiefly of carbon dioxide, with clouds rich in sulphuric acid. The axial rotation period is 243 days, which is longer than the revolution period (224.7 days), and radar investigations have revealed craters, mountains and deep canyons on the surface. Active vulcanism on Venus is a strong possibility.

There is one investigation which may be carried out this month. This concerns the so-called Ashen Light, or faint luminosity of the unlit side of Venus. The same sort of phenomenon in the case of the Moon is easily explained; it is due to light reflected on to the lunar surface from the Earth. But Venus has no satellite, and the existence of a night glow there is much less easy to account for. Yet it has been seen on countless occasions since it was first reported by Johann Hieronymus Schröter in the late 18th century.

Some astronomers are sceptical of its reality, and put it down to contrast effects, but others regard it as real. It could be due to electrical phenomena in the upper atmosphere of the planet—and we do now know that there is almost continuous lightning nearer the surface; but we have to admit that the Ashen Light is still very enigmatical.

Obviously, it may be observed only when Venus is a crescent, and the only safe procedure is to block out the bright crescent itself by means of an occulting bar fitted in the optical system of the telescope. If the Ashen Light remains visible even when the crescent is hidden, it may be regarded as a real phenomenon, but conclusive evidence is lacking at the present time.

Conditions in the latter part of December should be very favourable for searching for the Ashen Light. Telescopes of at least 6 inches aperture are desirable, and great care must be taken in making the observations; but this is a field of research in which the experienced amateur may be able to make a real contribution to our knowledge about this peculiarly hostile planet.

Finally, it may be emphasized again that Venus is not the Star of Bethlehem! December appearances comparable to that of 1981 are common enough, and would have excited no comment from the Wise Men or anyone else. Whether or not the Star of Bethlehem really appeared will probably never be known, but at any rate it cannot be put down to Venus.

Early in 1982, Venus will approach inferior conjunction, which it will reach on 21 January. It will then become a morning object, remaining so until the next superior conjunction on 24 November 1982; but all in all, Venus will not be so well placed for most of 1982 than it has been during the present year.

Eclipses in 1981

In 1981 there will be two eclipses of the Sun and two of the Moon.

(1) *A penumbral eclipse of the Moon,* on 20 January is unlikely to attract much attention, but on this occasion the Moon is totally immersed in the penumbral shadow of the Earth, and some change may be noticed. The eclipse may be observed generally in North and South America. Only the beginning will be seen in the British Isles, and the end in eastern Australia and New Zealand.

Eclipse begins $5^h 36^m$; maximum $7^h 50^m$; ends $10^h 04^m$.

(2) *An annular eclipse of the Sun* on 4 February, the central path beginning to the south of Australia and passing south of New Zealand to end near the west coast of Peru. Visible as a partial eclipse in New Zealand, the South Pacific Ocean and Antarctica.

At Dunedin, the eclipse begins at $19^h 42^m$, reaches its maximum (94 per cent) at $20^h 51^m$, and ends at $22^h 08^m$. Times at other parts of New Zealand will be much the same but the magnitude will be less. At Hobart, which is to the north of the central path, the eclipse begins at $19^h 38^m$, reaches a maximum of 98 per cent at $20^h 37^m$, and ends at $21^h 42^m$.

(3) *A partial eclipse of the Moon* on 17 July will have a magnitude of 55 per cent and will be visible in America. Only the beginning is visible in the British Isles, and the

end in New Zealand.
   Eclipse begins $3^h\ 35^m$; ends $6^h\ 09^m$.

(4) *A total eclipse of the Sun* on 31 July will have a maximum duration of $2^m\ 02^s$. The central path begins near the Black Sea and crosses the whole of the southern U.S.S.R. to end in the Pacific Ocean to the north of Hawaii. Visible as a partial eclipse over the greater part of Asia, the extreme north-west of North America and in Arctic regions.

All times are G.M.T.

Occultations in 1981

In the course of its journey round the sky each month, the Moon passes in front of all the stars in its path and the timing of these occultations is useful in fixing the position and motion of the Moon. The Moon's orbit is tilted at more than five degrees to the ecliptic, but it is not fixed in space. It twists steadily westwards at a rate of about twenty degrees a year, a complete revolution taking 18.6 years, during which time all the stars that lie within about six and a half degrees of the ecliptic will be occulted. The occultations of any one star continue month after month until the Moon's path has twisted away from the star but only a few of these occultations will be visible at any one place in hours of darkness.

Only four first-magnitude stars are near enough to the ecliptic to be occulted by the Moon; these are Regulus, Aldebaran, Spica, and Antares. The occultations of Aldebaran, which began in 1978, continue until March, but those of Regulus have now ceased. The planet Mars is occulted in the Southern Hemisphere on 5 February and on 24 September. Predictions of these occultations are made on a world-wide basis for all stars down to magnitude 7.5, but in recent years considerable attention has been paid to occultations of radio sources by the Moon, and of faint stars by satellites or minor planets. The exact timing of such events given valuable information about positions, sizes, orbits, atmospheres and some-times of the presence of satellites. The discovery of the rings of Uranus in 1977 was the unexpected result of the observations made of a predicted occultation of a faint star by Uranus. The duration of an occultation by a satellite or minor planet is quite small (of the order of a minute or less) and photoelectric timing is essential. If observations are made from a number of stations it is possible to deduce the size of the planet. A recent measure of the size of the minor planet Juno is described in the note on page 133. The final predictions for 1980 included times for occultations of stars by Neptune, Pluto and forty minor planets. A possible new measure of the diameter of Pluto promises to be of the greatest importance.

Comets in 1981

The appearance of a bright comet is a rare event which can never be predicted in advance, because this class of object travels round the Sun in an enormous orbit with a period which may well be many thousands of years. There are therefore no records of the previous appearances of these bodies, and we are unable to follow their wanderings through space.

Comets of short period, on the other hand, return at regular intervals, and attract a good deal of attention from astronomers. Unfortunately they are all faint objects, and are recovered and followed by photographic methods using large telescopes. Most of these short-period comets travel in orbits of small inclination which reach out to the orbit of Jupiter, and it is this planet which is mainly responsible for the severe perturbations which many of these comets undergo. Unlike the planets, comets may be seen in any part of the sky, but since their distances from the Earth are similar to those of the planets their apparent movements in the sky are also somewhat similar, and some of them may be followed for long periods of time.

The number of comets under observation in any one year is much greater than is generally supposed. The following table gives the numbers of newly discovered comets, of periodic comets recovered as a result of successful predictions, and of comets followed from previous years.

	1975	1976	1977	1978
New discoveries	13	5	8	11
Predicted and recovered	4	6	11	7
Still under observation	14	16	13	17
	31	27	32	35

Data for 1979 are not yet complete, but seven new comets were discovered and five more were recovered as a result of accurate predictions of their returns. The following comets are expected to return to perihelion during 1981:

Comet Schwassmann-Wachmann (2) was discovered in 1929 and made its eighth appearance in 1973-1975, when it was widely observed over a long period. The resulting accurate prediction enabled observers to recover the comet at the end of 1979, and it is expected at perihelion in March 1981. The orbit of this comet lies entirely outside that of Mars, but is only moderately eccentric and has a period of 6.5 years.

Comet Reinmuth (2) was discovered in 1947 and made its fifth appearance in 1973. Its orbit also has a period of 6.5 years and lies entirely outside the orbit of Mars.

Comet Borrelly was well observed in 1973-1975, its ninth appearance. It was first seen in 1905, and has a period of 6.8 years and the rather large inclination of 30 degrees. It is expected at perihelion in February.

Comet West-Kohoutek-Ikemura was first detected by R. J. West in January 1975 when examining a plate taken the previous October at the European Southern Observatory of La Silla, Chile. The comet was recovered accidentally by Kohoutek at Hamburg in February and by Ikemura in Japan. The orbit has a period of 6.1 years and an inclination of 30 degrees.

Comet Kohoutek, discovered in 1975, was the fifth cometary discovery made by Kohoutek at Hamburg-Bergedorf. This is a typical short-period comet, with a small inclination, moderate eccentricity and a period of 6.2 years.

Comet Finlay has been known since 1886. It has not been seen at each return to the Sun, but it has made nine appearances in all, the last in 1974. It has a period of nearly seven years, and is expected at perihelion in June.

Comet Gehrels (2) was found in 1973 at the Hale Observatory on Palomar Mountain. The comet has a period of nearly eight years and should return to the Sun at the end of 1981.

Comet Swift-Gehrels was the third comet discovered by Gehrels at Palomar in 1973. It was only a faint object and only a few observations were made, but these were sufficient to show that this comet was identical with Comet Swift (1) which had appeared only once before in 1889-90. The orbit showed very little change in the nine revolutions it had made since 1889. The period is 9.2 years, and the orbit reaches out well beyond the orbit of Jupiter.

Comet Longmore was discovered in June 1975 by A. J. Longmore at Siding Spring Observatory, N.S.W. The comet had passed perihelion in the previous November, and very few observations were obtained. This comet seems to have been seriously perturbed by Jupiter in 1963, and it now has a period of nearly seven years and an inclination of 24 degrees.

Comet Slaughter-Burnham was found at Lowell, Arizona, in 1958 and made another appearance in 1970. This comet has the longer period of 11.6 years but only a moderate eccentricity.

Comet Kearns-Kwee was discovered in 1963 at Palomar and appeared again in 1971-2. The comet made a close approach to Jupiter in 1961 and its period was reduced from 51 years to 9 years. It is expected to return to perihelion at the end of November.

A number of comets having small orbits or moderate eccentricity can be observed each year, generally at the time of opposition. Among these are three which are regularly observed:

Comet Encke has the smallest known orbit, with a period of only 3.3 years. The orbit is very eccentric, but with modern large telescopes it can be observed even at aphelion.

Comet Gunn was discovered at Palomar in 1970 and has a period of 6.8 years.

Comet Schwassmann-Wachmann (1) has a large orbit of small eccentricity which lies entirely between the orbits of Jupiter and Saturn. The period is 16 years, but it is observed every year, and is remarkable for its outbursts of brightness. Normally of magnitude +18, it may suddenly brighten by several magnitudes, and has been known to be as bright as magnitude +10.7.

Meteors in 1981

Meteors ('shooting stars') may be seen on any clear moonless night, but on certain nights of the year their number increases noticeably. This occurs when the Earth chances to intersect a concentration of meteoric dust moving in an orbit around the Sun. If the dust is well spread out in space, the resulting shower of meteors may last for several days. The word 'shower' must not be misinterpreted—only on very rare occasions have the meteors been so numerous as to resemble snowflakes falling.

If the meteor tracks are marked on a star map and traced backwards, a number of them will be found to intersect in a point (or a small area of the sky) which marks the radiant of the shower. This gives the direction from which the meteors have come.

The following table gives some of the more easily observed showers with their radiants; interference by moonlight is shown by the letter M.

Limiting dates	Shower	Maximum	R.A.	Dec.	
Jan. 1-6	Quadrantids	Jan. 4	$15^h 28^m$	$+50°$	
Mar. 14-18	Corona Australids	Mar. 16	$16^h 20^m$	$-48°$	
April 20-22	Lyrids	April 21	$18^h 08^m$	$+32°$	M
May 1-8	Aquarids	May 5	$22^h 24^m$	$+00°$	
June 17-26	Ophiuchids	June 20	$17^h 20^m$	$-20°$	M
July 15-Aug. 15	Delta Aquarids	July 25	$22^h 36^m$	$-17°$	
July 15-Aug. 20	Pisces Australids	July 31	$22^h 40^m$	$-30°$	
July 15-Aug. 25	Capricornids	Aug. 2	$20^h 36^m$	$-10°$	
July 27-Aug. 17	Perseids	Aug. 12	$3^h 04^m$	$+58°$	
Oct. 15-25	Orionids	Oct. 20	$6^h 24^m$	$+15°$	
Oct. 26-Nov. 16	Taurids	Nov. 8	$3^h 44^m$	$+14°$	M
Nov. 15-19	Leonids	Nov. 17	$10^h 08^m$	$+22°$	
Dec. 9-14	Geminids	Dec. 14	$7^h 28^m$	$+32°$	M
Dec. 17-24	Ursids	Dec. 22	$14^h 28^m$	$+76°$	

M=moonlight interferes

Minor Planets in 1981

Although many thousands of minor planets (asteroids) are known to exist, only 2,100 of these have well-determined orbits and are listed in the catalogues. Most of these orbits lie entirely between the orbits of Mars and Jupiter. All of these bodies are quite small, and even the largest can only be a few hundred miles in diameter. Thus, they are necessarily faint objects, and although a number of them are within the reach of a small telescope few of them ever reach any considerable brightness. Of these the most important are the 'big four', Ceres, Pallas, Juno and Vesta. Vesta can occasionally be seen with the naked eye, and this is most likely to occur at a June opposition, when Vesta is at perihelion. In 1981, Vesta is at opposition on 21 February, magnitude 6.2, and it will then be in Leo near the star Gamma Leonis. Ceres reaches magnitude 6.5 when at opposition on 10 January in Gemini. The planet is then to the west of Castor and Pollux. There is no opposition of Pallas during the year, and when at its brightest in December, it only reaches magnitude 8.5. Juno is even fainter at magnitude 10.1 when at opposition in Virgo on 20 April.

A vigorous campaign for observing the occultations of stars by the minor planets has produced new figures for the dimensions of some of them, as well as the suggestion that some of these planets may be accompanied by satellites. Some results for 1979 have already been announced, and these include a preliminary value for the size of Juno. This planet occulted a ninth magnitude star on 11 December 1979, and accurately timed records were made at eleven stations. Assuming an elliptical outline for the planet, the axes of the ellipse were found to have lengths of 291 km and 243 km. If these figures are confirmed, Juno would appear to be rather larger than we had previously believed. There was no evidence of the existence of any satellites.

Some Events in 1982

In 1982, there will be seven eclipses—the maximum number possible.

9 January—total eclipse of the Moon—Asia, Australasia, Europe, Africa.

25 January—partial eclipse of the Sun—Antarctica, southern part of New Zealand.

21 June—partial eclipse of the Sun—extreme south of Africa.

6 July—total eclipse of the Moon—America, Australasia.

20 July—partial eclipse of the Sun—Arctic regions, northwest of Europe and North America.

15 December—partial eclipse of the Sun—Europe, western Asia.

30 December—total eclipse of the Moon—America, Australasia, Asia.

THE PLANETS

Mercury may best be seen in northern latitudes about the time of eastern elongation (9 May, evening star) and of western elongation (morning star) on 17 October. In the Southern Hemisphere the most favourable times are about 26 February (morning star) and 6 September (everning star).

Venus is in inferior conjunction in January, and then becomes a morning star, reaching greatest western elongation on 1 April and superior conjunction in November.

Mars is at opposition on 31 March in Virgo.

Jupiter is at opposition on 26 April near the borders of Virgo and Libra.

Saturn is at opposition on 9 April in Virgo.

Uranus is at opposition on 24 May in Scorpius.

Neptune is at opposition on 17 June in Ophiuchus near the border with Sagittarius.

Pluto is at opposition on 15 April in Virgo near the border with Boötes.

PART TWO

Article Section

The Quadrantid Meteor Shower

JOHN MASON

Almost everybody has, at one time or another, seen a meteor. The brief streak of light in the night sky, lasting but a few tenths of a second, is merely the spectacular demise of a tiny dust particle. Prior to its encounter with the Earth, the meteoroid, as it is known, is minding its own business, pursuing an orbit around the Sun. However, once captured by the Earth's gravitational field, the meteoroid enters the upper atmosphere and is rapidly heated to incandescence, through friction with the atmospheric molecules. The majority of meteoroid particles are very small, no bigger than a grain of sand, and one the size of a grape would produce a very bright meteor indeed. This is hardly surprising when one realizes that all meteors are travelling at a speed of between 11 and 72 km/sec relative to the Earth, at the time of encounter. The brightest meteors, which have a luminosity that exceeds that of the brightest planets, are known as fireballs or bolides.

At least 80 per cent of all the meteors seen in the night sky are what are called 'sporadic' meteors. Countless millions of these bodies are found between the major planets, each one following an individual orbit in space. On any clear, moonless night, a visual observer will see between 5 and 15 of these sporadic meteors, every hour. The rate actually varies with the time of day (peak rates occurring at about 4 a.m.), and from month to month, throughout the year. At certain times of the year, the Earth's orbit happens to pass through narrow regions where the meteoric dust is far more concentrated, and the number of meteors visible increases sharply. These meteor streams, as they are known, are formed from clouds of dust particles left behind by a periodic comet along its orbit. The stream particles may be concentrated in one region of the comet's orbit, or they may be spread out uniformly around it and each year, when the Earth intersects the comet's orbit, a meteor shower will be seen, lasting from a few hours to several days, depending on the width of the meteor stream. One of the most well-known streams is that due to Comet Swift-Tuttle of 1862, and this produces the Perseid meteor shower, visible every year during the first two weeks of August. The meteoroids in a stream, all move along parallel paths, but to an observer on Earth, perspective will make it appear that all the meteors are emanating from one area of the sky, known as the radiant of the meteors. It is usual practice to name a meteor stream after the constellation in which the radiant lies. Thus, the Perseids radiate from a point in Perseus.

One of the most interesting meteor showers is visible during the first week of January. This stream, the Quadrantids, is certainly one of the most difficult to cover during the year, and it has a notoriously bad name with meteor observers. The shower occurs in the depths of winter, and one needs the characteristics of a polar bear to observe it, visually, at all. Furthermore, heavy cloud cover is often the 'order of the day' in early January, and when it is clear, you can be certain that there will be a bright Moon to drown out all but the brightest members of the stream! A further problem with the Quadrantids is that the stream of meteors, which produces the shower as the Earth passes through it, is extremely narrow (the activity only exceeds one-quarter of the maximum rate, for 19 hours), and very often, maximum activity occurs during the hours of daylight, from a particular observing site. In addition, the altitude of the radiant is always low during evening hours, and a respectable altitude is not attained until after 2 a.m., with dawn occurring around 7 a.m. The low altitude vastly reduces the number of meteors seen by a visual observer, and so true rates are only noted during the early morning hours. All these factors conspire to ruin any planned observations, and, probably, on only one occasion in every ten years, can the Quadrantid maximum be observed under ideal conditions, that is, where part of the 19-hour period occurs between 2 a.m. and 7 a.m. Nevertheless, the visual activity of the Quadrantids is, at present, the greatest of all the annual meteor showers. On a fine, moonless night, an observer may see over 120 meteors per hour (120 m/h) at maximum, and rates as high as 200 per hour were recorded in 1965. It is this fact which lures visual meteor observers out into the cold and, often, snowy landscape, to monitor the shower's activity. Some North American observers have to cope with temperatures of up to –40°C (also –40°F), and build special observing shelters to aid their work!

The story behind the name of the shower, the Quadrantids, is an interesting one. In 1930, the International Astronomical Union standardized the boundaries of all the constellations, numbering 88 in total. Originally, constellations had no boundaries, and in the 17th and 18th centuries, the astronomers of the time added various confusing constellation regions of their own choice, often with clumsy names, the majority of which are now obsolete. The first boundaries were drawn up by Bode, in 1801, and one of the constellations in his list was Quadrans Muralis (the Mural Quadrant), located exactly where the radiant of the January meteors lay. It is likely that the area was introduced by J. J. Lalande, around 1795, to commemorate the instrument used to observe the stars listed in his catalogue. The name, Quadrans, has now been rejected, and the radiant point of the shower now lies in the constellation of Boötes, mid-way between the stars Beta Boötis and Theta Draconis, in fact, not far from the tail of Ursa Major.

Although, today, the Quadrantids produce a visual shower of high activity, this is only a comparatively recent state of affairs. The comet which originated the shower is not known, as the Quadrantids do not follow the orbit of any comet observable at the present. We can, therefore, conclude that the parent comet has probably disintegrated, rather like Comet Biela, which split up in 1846, and was not seen after 1852. The Quadrantids are certainly a very young stream, having a large proportion of small particles concentrated in a narrow region. Historical records show that it is quite likely that the Earth never intersected the Quadrantid stream before the early part of the 19th century. The annual occurrence of the January shower was noted by Wartmann, in Geneva, from 1835-8, and, independently described by A. Quetelet, at Brussels, in the first general catalogue of meteor showers (1839), and by E. C. Herrick, in the United States, during 1836-7. Without doubt, the first date appears to be 1 January 1835, with little or no activity before then. The radiant point of the shower was first accurately determined, in January 1863, by Stillman Masters, in America, and data on the shower are available, almost continuously, from that date.

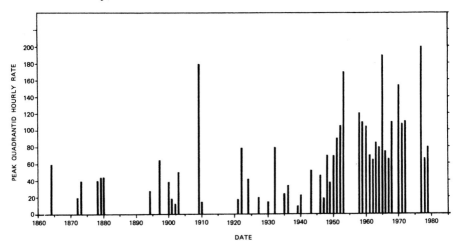

Figure 1. The fluctuation in peak Quadrantid rates, from 1860 to date, is clearly shown here. Note especially, the consistently high rates of the last 30 years.

The first major summary of Quadrantid activity was made in 1930, by Willard J. Fisher. He examined the observed Quadrantid rates between 1863 and 1927, and much of our knowledge of the early history of the shower is derived from his work. Particularly high rates were noted in 1864 (60 m/h), 1897 (65 m/h) and 1909 (180 m/h), the latter display being unequalled until the last thirty years. The work of J. P. M. Prentice was especially useful between 1921 and 1953, in which he examined the peak rates of many Quadrantid displays. He showed that during the first half of this century, an average display produced about 45 m/h, with a steady rise to maximum, followed by a rapid fall. In 1953, there was another strong return, producing 170 m/h, and since 1965, exceptionally high activity has become the norm for this shower. An average rate of 115 m/h was observed over the six years between 1965 and 1971, and this trend was continued for the rest of the 1970s.

The display of 1977 seems to have been particularly noteworthy, although the Moon was only two days from Full at the time. Maximum was predicted for 4 p.m. on 3 January, and observers, before dawn on that day, reported that Quadrantids were exceptionally numerous, with hourly rates variously estimated at up to 200 m/h at peak, and possibly greater. On one occasion, 12 meteors were noted in one minute. Many astronomers thought that the meteors seemed to come in short bursts of high activity. The sky was certainly extremely clear on this occasion, and several bright meteors were seen, including magnitude –5 and magnitude –12 fireballs, the latter rivalling the Full Moon in brilliance. In 1978, conditions were generally poor, especially after midnight on 3/4 January, but an hourly

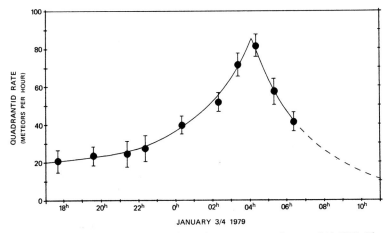

Figure 2. The variation in visual Quadrantid meteor rates on January 3/4, 1979. The vertical bars indicate the probable errors on each point. Note the sharp peak, and the rapid fall from maximum. The dotted line indicates that daylight prevented further observation from the British Isles.

143

rate of at least 65 m/h occurred at maximum, although the radiant altitude was extremely low at this time. Observers of the 1979 Quadrantid display, in Great Britain, had to cope with some of the worst weather conditions this century. Heavy snow had fallen over the New Year, and, on the night of 3/4 January, there were several inches of snow on the ground, and the temperature was –6°C! Insulated sleeping bags, hot water bottles and well muffled ears and fingers were a necessity. The night was clear, and Quadrantid rates rose from 25 m/h at 10 p.m. on the 3rd to a peak of around 80 m/h at 4 a.m. on the 4th. The conditions in 1980 could not have been worse, with maximum occurring in broad daylight and a Full Moon, high up in Gemini. Nevertheless, some Quadrantids were seen on 4 January.

The considerable fluctuation in the peak Quadrantid rates shown in recent years, is almost certainly due to the fact that the Earth samples a different part of the meteor stream each January. In a young meteor stream, such as the Quadrantids, there are wide variations in meteoroid density across the stream. There will be diffuse regions of older material at the outer edge, surrounding a dense, compact central swarm, which is the youngest matter. It is worth noting that, although peak rates of 300 m/h are possible if the Earth chances to intersect the core of the stream, there is no possibility of a future Quadrantid 'storm', in which many thousands of meteors per hour are seen. This was the case in 1966, with the Leonid swarm, due to Comet Tempel-Tuttle of 1866, when observers in Arizona noted rates of over 100,000 m/h in the early morning of 17 November. For this to happen, the associated comet must still be laying down meteoroids into the stream, forming intensely concentrated filaments in the core. As the comet associated with the Quadrantids is unknown, and has not been observed, a storm is very unlikely.

Observations of the Quadrantid shower during the last 30 years, have given us great insight into the structure of the meteoroid stream which causes it. Up to 1950, astronomers were largely dependent on the visual and photographic work of amateurs, but since that date, observations by radar, high

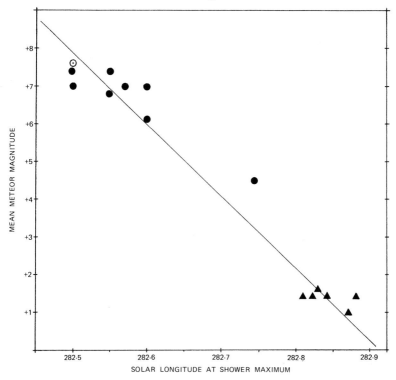

Figure 3. The time of Quadrantid maximum, denoted here by the solar longitude, varies for meteoroids of differing brightness. Peak rates for the smallest (and fainter) particles, detected by telescopic (⊙) and radar (●) means, occur up to 7 hours earlier than for the brighter (and larger) meteoroids observed visually and photographically (▲).

sensitivity television systems, coupled with telescopic work, have vastly improved the coverage. A visual observer, using the naked eye, will see only the brightest meteors, but radar will pick up the tiny particles which are well below the visual threshold and, of course, it can operate in a cloudy sky. A comparison between the Quadrantid rates obtained by visual and shortwave radio methods has produced an interesting result, and recent inclusion of telescopic meteor observations has substantiated the conclusion. It seems that at the fainter magnitudes observed telescopically, the Quadrantid peak occurs almost 7 hours before visual maximum. This seems to suggest that the meteoroids in the swarm have been dispersed according to their mass, such that, as the Earth moves through the stream it meets the smaller and, hence, the fainter particles first. In every case, including the radar data, the maximum for faint meteors occurs earlier than for bright. Over the brightness range, between magnitudes $+1$ and $+8$, the Quadrantid peak occurs about $1\frac{1}{4}$ hours earlier for every magnitude step towards the fainter particles. Comparison of photographic and radar data has also shown that the fainter Quadrantids have shorter orbital periods (4.7 years) compared to the brighter ones (5.2 years). This, again, shows that the particles are dispersed according to their mass.

Confirmation that the Quadrantid swarm consists of diffuse and compact components has been obtained by an examination of the radiant itself, during the period of the shower's activity. During the early and late stages of the shower, the radiant area is complex and may be over 10 degrees across, with several separate sub-centres of radiation. However, at the time of peak rates, the radiant contracts, becoming a sharp, radiating area, less than a degree across. The Earth first encounters the scattered outer meteoroids, then passes through the dense centre, before moving outwards, through the fringe once more.

If one looks at the distribution of the brightness of the Quadrantid particles as they appear on entry into the atmosphere, and compares this with the sporadic background of meteor activity, present at all times throughout the year, we obtain some indication of the spread of meteoroid sizes in the swarm itself. One finds that, for the fainter meteors, the shower is deficient compared to the sporadics, but for the brightest particles the Quadrantids are slightly richer. This effect is very pronounced for the Perseid stream, visible every August, and so we may deduce that the Quadrantid swarm is much the younger of the two. This is because as soon as any meteor stream is formed, tiny perturbations cause the steady loss of the smallest particles, and as time goes on, the stream becomes relatively richer in the large ones.

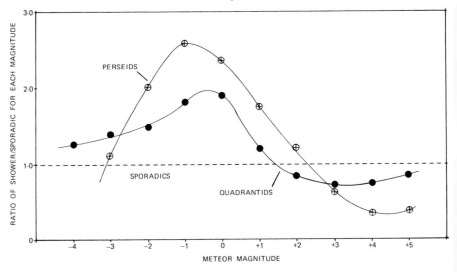

Figure 4. If shower meteors had the same magnitude and, hence, meteoroid size distribution, as sporadic meteors, all points would lie along the horizontal dotted line. The Perseid stream is far richer in bright meteors, and deficient in faint ones, relative to sporadics. For the Quadrantids, this effect is not so marked, showing it to be the younger of the two streams.

The final clues to the history, and current and future activity of the Quadrantids, are supplied when one examines the actual orbits of the meteoroids around the Sun. The meteor stream, at present, has a high orbital inclination of 71°.4, when compared with short period comets and the other major meteor streams. If this high inclination had always been present, then we should not observe the dispersion in the meteoroid masses inferred from the comparison of the visual and radar data. The reason for this is that the segregation of the radar meteoroids from the visual ones is due to a phenomenon known as the Poynting-Robertson effect, and this will only occur in the plane of the stream's orbit, and not the Earth's orbit, as is observed to be the case. We, therefore, deduce that in the past, the Quadrantids had a completely different orbit, with a low orbital inclination, and that the mass dispersion took place at that time. Recent work by Hughes, Williams and Murray has shown that, about 1,500 years ago, the orbital inclination was only 12 degrees,

147

and that the Quadrantid stream itself was formed by a periodic comet with a period of about 5.4 years, which broke up between 1,700 and 1,300 years ago. The meteoroid particles ejected, quickly spread around the comet's orbit in a period of 12 to 35 years, before the orbital inclination slowly increased to its present-day value. This time, for the particles to be distributed around the orbit is known as the loop formation time, and it was so short for the Quadrantids because the orbit at that time, passed within only 13 million kilometres of the Sun at its closest (known as the perihelion distance), compared to the present-day value of about 146 million kilometres.

The Quadrantid orbit is also moving in space, relative to the major planets of the Solar System, due to gravitational perturbations, particularly by the giant planet Jupiter. Some of the closest encounters with this planet have had a major effect on the stream, and Jovian perturbations are probably the main influences governing the activity of any given display. The orbit may be shifted slightly from side to side, causing the Earth to intersect only the outer, less dense, regions, leading to a weaker display, as was the case during the first half of this century. However, in recent years, the Earth must have been passing closer to the stream centre, resulting in the currently high rates, In fact, gravitational perturbations introduce a slow change in all the orbital parameters of the meteoroid swarm. The position of the ascending node of the Quadrantid orbit is easy to measure, as it depends on the time of peak shower activity. At present, the exact position of the node is retrogressing steadily, at a rate of about $0°.4$ century. This is very slow, but its effect is to cause the time of the Quadrantid maximum to occur earlier each year, at a rate of about $6\frac{1}{2}$ hours per century.

When we consider what will happen to the stream's orbit in the future, some surprising results are obtained. Every 11.86 years, Jupiter passes very close to the aphelion area of the stream, and perturbs a considerable proportion of the particles. Some meteoroids have their orbits drastically changed, and are completely removed from the swarm. For others, the consequences are less severe and the effects of the close encounter are detected about two years afterwards, when these meteors have progressed around the orbit to the vicinity of the Earth. The perturbations tend to change the mean ascending node of the stream (and hence the time of maximum) about every 12 years. It is interesting to note that, on those occasions, when Jupiter passes very close to the stream, there may well be a meteor shower in the atmosphere of this planet. As the perihelion distance of the meteor stream increases in the future, it will become impossible for the Earth to intersect it and, therefore, it is highly likely that this, coupled with the future dispersion of the meteoroid particles, will mean that the Quadrantid meteor stream will cease to be with us about 500 years from now. The Quadrantids, therefore, present us with a

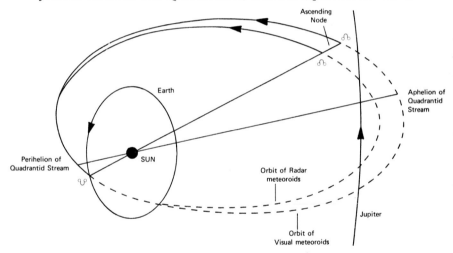

Figure 5. The orbits of the Quadrantid meteor stream particles and their relationship with the orbits of the Earth and Jupiter. The brighter, visual meteoroids move in larger orbits than the fainter, radar ones. The Earth is at ♈ on January 3 each year.

golden opportunity to monitor the evolution of a meteor stream in a fairly short space of time. Observations of the Quadrantids are always required, and amateurs can carry out useful visual and photographic work with a little resourcefulness, to overcome the atrocious weather conditions normally present in early January.

In 1981, the maximum for visual Quadrantid meteors will occur at sunset (4 p.m.) on Saturday, 3 January. This is one of the most favourable opportunities to observe the shower for many years, and should not be missed. Not only does the peak occur at a weekend, when observers can catch up on their sleep during the day, but the Moon will be 27 days old, almost New, and will not interfere with the shower at all. Very high rates may be observed before dawn on 3 January. Let us hope we have a similar display to that of 1977, and a clear sky! Observations should also be carried out on the evening of 3 January, even though the radiant will be rather low.

References

Alcock, G. E. D., and Prentice, J. P. M., 1953. *J. Brit. astr. Assoc.*, **63**, 186.

Backhouse, T. W., 1884. *Astr. Register*, **22**, 16.

Belkovich, D. I., and Tohktasev, V. S., 1974. *Bull. Astron. Inst. Czech.*, **25**, 370.

Chambers, G. F., *Handbook of Astronomy*, 117, 623, 640. Oxford, Clarendon Press, 1890.

Denning, W. F., 1888. *Mon. Not. R. astr. Soc.*, **48**, 110.

—1899. *Mem. R. astr. Soc.*, 53, 264.

Fisher, W. J., *Circ. Harvard College Obs.*, No. 346 (1930).

Hamid, S. E., and Youssef, M. N., 1963. *Smithson. Contr. Astrophys.*, **7**, 309.

Hawkins, G. S., and Almond, M., 1952. *Mon. Not. R. astr. Soc.*, **112**, 219.

Hindley, K. B., 1968. *J. Brit. astr. Assoc.*, **79**, 68.

—1970. *J. Brit. astr. Assoc.*, **80**, 479.

—1971. *J. Brit. astr. Assoc.*, **82**, 57.

Hoffmeister, C., *Meteorströme*, Weimar, 1948.

Hughes, D. W., 1972. *Observatory*, **92**, 41.

—1974, *Mon. Not. R. astr. Soc.*, **166**, 340.

—1974. *Space Res.*, XIV, 709.

—1977. *Space Res.*, XVII, 565.

Hughes, D. W., and Taylor, I. W., 1977. *Mon. Not. R. astr. Soc.*, **181**, 517.

McIntosh, B. A., *Comets, Asteroids, Meteorites., Interrelations, Evolution and Origins.*, Ed. by A. H. Delsemme, 171, Univ. of Toledo Press, 1977.

Millman, P. M., and McKinley, D. W. R., 1953. *J. Roy. astr. Soc.*, Canada, **47**, 237.

Poole, L. M. G., Hughes, D. W., and Kaiser, T. R., 1972. *Mon. Not. R. astr. Soc.*, **156**, 223.

Porter, J. G., *Comets and Meteor Streams*, Chapman & Hall Ltd., 1952.

Prentice, J. P. M., 1940. *J. Brit. astr. Assoc.*, **51**, 19.

—1953. *J. Brit. astr. Assoc.*, **63**, 175.

Quetelet, A., *Catalogue des Principales Apparitions d'Etoiles Filantes.*, Brussels, 1839.

—*Physique du Globe*, 290, Brussels, 1861.

Rigollet, R., 1951. *Ann. d'Astrophys.*, **14**, 181.

Williams, I. P., Murray, C. D., and Hughes, D. W., 1979. *Mon. Not. R. astr. Soc.*, **189**, 483.

—1979. *Mon. Not. R. astr. Soc.*, **189**, 493.

Ariel VI: A British Satellite for High Energy Astronomy

J. L. CULHANE

Until the middle of this century astronomical observations were undertaken with telescopes on the surface of the Earth and were therefore limited to those radiations that were able to penetrate the Earth's atmosphere. These included visible or white light together with radio waves and some infra-red radiation. However, it is clear from the spectral chart (Figure 1) that the electromagnetic spectrum extends beyond what we know as visible light to ultra-violet, X-rays and gamma rays. These latter radiations are absorbed by the gases of the atmosphere and are thus unable to penetrate to the Earth's surface. In addition, there are high energy charged particles, confusingly known as cosmic rays, though they are quite different from electromagnetic radiation, which reach the top of the atmosphere but are also unable to penetrate to ground level.

Detailed studies of all of these phenomena first became possible with the availability of space vehicles. At first, only sounding rockets were employed. These were capable of reaching heights greater than 100 km above the surface of the Earth, but could only remain at these altitudes for a few minutes before falling back into the atmosphere. However, with the launching of Sputnik in 1957, the capability to place satellite vehicles in permanent orbit around the Earth at heights in excess of 500 km became available. Such satellites could carry instrumentation above the Earth's atmosphere for periods of many years and thus for the first time the detailed study of high energy radiation and particles became possible.

In the period from 1962 to the present, the United Kingdom

The Electromagnetic Spectrum

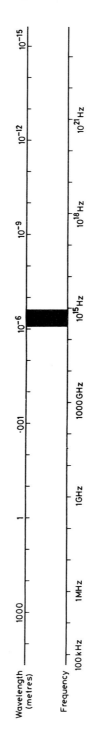

Radio	Millimetre	Infra red	Light	Ultra violet	X-rays		Gamma-rays

Wavelength (metres)

10^{-15}	10^{-12}	10^{-9}	10^{-6}	.001	1	1000

Frequency

100kHz 1MHz 1GHz 1000GHz 10^{15}Hz 10^{18}Hz 10^{21}Hz

long wave

short wave   television   radar   lamps   causes sunburn   radiation treatment e.g. cobalt bomb
VHF   microwave ovens   heaters   causes fluorescence   sterilization of equipment
microwave links   disco lights

Figure 1. A chart of the electromagnetic spectrum from gamma rays to radio waves.

has constructed and launched a number of satellites for scientific research. We will briefly discuss the nature of these spacecraft and, in particular, the fifth member of the series, Ariel V, which was devoted to X-ray astronomy. We will then describe the most recently launched spacecraft in the series, Ariel VI. Although there has been some interference with the operation of this satellite by ground based transmissions, it is operating successfully and returning useful scientific data. It carries three experiments for the study of cosmic rays and X-rays. These three instruments will be described and some of the early results obtained with them will then be presented.

The United Kingdom Ariel Series of Scientific Satellites

Prior to Ariel VI, five other UK satellites were launched in a co-operative programme with the US National Aeronautics and Space Administration that began with the flight of Ariel I in 1962.

Details of these spacecraft, their launch dates and their payloads are given in Table 1. Ariel VI will be discussed in more detail later. Of the previous spacecraft in the series, only Ariel I and Ariel V have undertaken major investigations in astronomy. The X-ray instrument on Ariel I demonstrated for the first time the dramatic increase in the temperature of a small part of the solar corona during the occurrence of a solar flare. It also established that the corona above active regions contained high temperature material. The interaction of solar X-rays with the Earth's ionosphere was also studied.

The Ariel V satellite was entirely devoted to cosmic X-ray astronomy, and was undoubtedly the most successful of the five spacecraft listed in Table 1. The University College X-ray spectrometer was the first to detect X-ray emission lines from a number of astronomical objects (see Figure 2). The most important of these observations was undoubtedly the discovery of an iron emission line in the spectrum of the Perseus cluster of galaxies. Previously, it was not clear whether the extended X-ray sources associated with clusters were due to the presence of large volumes of high temperature ($T \sim 10^8$ K) gas or to some other non-thermal mechanism. Detection of the

TABLE 1 — **The Ariel Series of Scientific Satellites**

Spacecraft	Launch Date	PAYLOAD	
		Investigation	Scientific Group/Agency
Ariel I	26 April, 1962	Solar X-ray studies	University College London & Leicester University.
		Solar Lyman-α flux	University College London
		Ionospheric Composition & temperature	University College London
		Cosmic Ray Particle Flux	Imperial College
Ariel II	27 March, 1964	Galactic Radio Noise	Cambridge Univ.
		Ozone Distribution	Met. Office
		Micrometeorites	Manchester Univ.
Ariel III	5 May, 1967	Ionospheric Electron Density	Birmingham Univ.
		Galactic Radio Noise	Manchester Univ.
		Very low Frequency Radiation	Sheffield Univ.
		Ozone Distribution	Met. Office
		Terrestrial Radio Noise	Radio & Space Research Station
Ariel IV	11 Dec, 1971	Wave Particle Interactions	Birmingham, Manchester, Sheffield & Iowa Universities & Radio and Space Research Station.
Ariel V	15 Oct, 1974	X-ray Sky Survey	Leicester Univ.
		X-ray Spectra 1-20 keV	University College London
		X-ray Source Positions	University College London & Birmingham University.
		X-ray Spectra 15-500 keV	Imperial College
		All Sky X-ray Camera	Goddard Space Flight Center

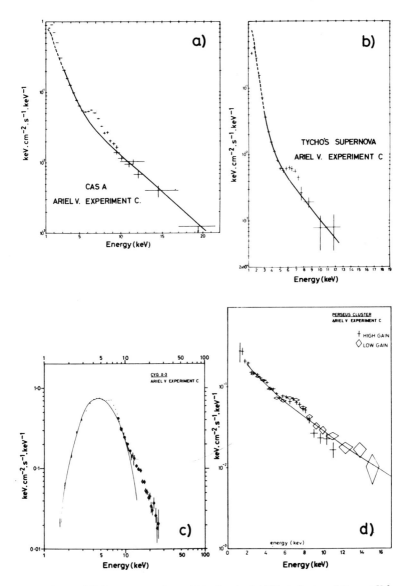

Figure 2. Ariel V X-ray spectra from a) Cassiopeia A, b) Tycho's star, c) Cygnus X-3, d) Perseus cluster indicating the presence of iron emission lines.

iron line firmly established the presence of hot gas and eliminated the non-thermal mechanisms from further consideration. The discovery of emission lines in the X-ray spectra of supernova remnants also established the presence of high temperature gas in these objects and demonstrated the absence of non-thermal pulsar-generated radiation in cases other than that of the Crab Nebula.

The spectrometer also detected the first slowly rotating neutron star. Prior to this observation, radio pulsars had been shown to be spinning neutron stars with periods of a few seconds or less. Ariel V found a neutron star in a binary system with a period of 6.7 minutes (Figure 3). Many other slow rotators were soon identified both by Ariel V and by other spacecraft. Their discovery has increased our understanding of neutron stars and of the evolution of stars in binary pairs. We will later discuss Ariel VI observations of these systems. The mapping instrument on Ariel V discovered many short-lived or transient X-ray sources. Indeed, the first slow rotator to be discovered was associated with a transient source in Centaurus. At least one of these transients was later identified with an optical nova. Detailed maps were made of the galactic centre region and several new sources were discovered in this area of the galaxy. Finally, the sky survey instrument, in the course of discovering many new extra-galactic X-ray sources, established Seyfert galaxies as an important class of X-ray emitting object.

Ariel V survived in orbit for more than five years and indeed was still returning useful data when Ariel VI, the subject of the present article was launched. Together with the UK X-ray telescope on the NASA Copernicus satellite Ariel V has established UK astronomy in a position of world leadership during the seventies.

The Ariel VI Spacecraft and its Experiments

The spacecraft together with its complement of experiments is illustrated in Figure 4. The structural design is based on that of the Ariel IV satellite but in this case the spherical cosmic ray detector is mounted on top of the cylindrical body. The two X-

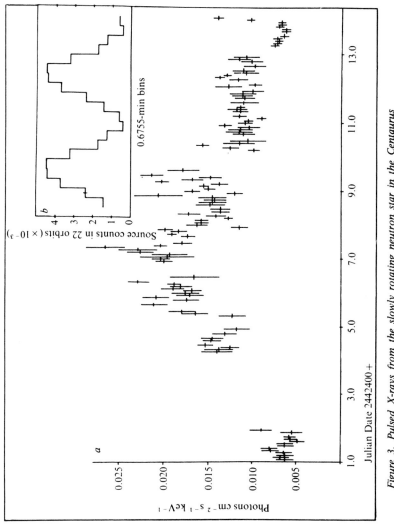

Figure 3. Pulsed X-rays from the slowly rotating neutron star in the Centaurus transient source as observed by Ariel V.

ray experiments are mounted around the outside of the body at four locations 90 degrees apart.

In orbit, the spacecraft spins about an axis which passes through the top of the spherical detector and down the centre of the cylindrical body. After launch, the four solar panels are deployed as shown in the figure. The base of the satellite and the solar array point in the direction of the Sun. In fact, a line perpendicular to the base is not permitted to be more than 60 degrees away from the direction of the Sun, otherwise the array would not develop sufficient power for the needs of the satellite. With the base pointed to Sun, the cosmic ray detector is shadowed by the body and can therefore remain at a lower temperature than other parts of the satellite. The two X-ray experiments also look away from the Sun, which is itself a strong X-ray source whose emission could hinder the observation of the much fainter cosmic sources.

A ring around the base of the satellite contains a large coil. When this is supplied with current, the interaction of this current with the Earth's magnetic field causes the spin axis of the vehicle to move around the sky. By controlling the strength of the current and the times at which it is switched on, the spacecraft spin axis can be pointed anywhere in the sky within the allowed limits. Thus, it is possible to study a large number of X-ray sources during the two-year mission lifetime.

Power from the solar cells is used to charge a battery which in turn supplies regulated power to experiments and spacecraft sub-systems. Data gathered by the experiments can be transmitted directly to a ground station when the satellite is overhead. However, during the long periods for which the satellite is out of sight of the primary ground station, data are stored on two tape recorders and are played back through the transmitters when the satellite is again over the ground station. In addition, the experiments themselves are equipped with buffer memories to enable them to occasionally acquire data with very good time resolution. These memories can also be used as data stores in the event of tape recorder failure. In addition to its data recording and transmitting systems, the satellite contains a command receiver which can act on instructions

Figure 4. An illustration of the Ariel VI satellite showing the principal features of its construction and the location of the experiments.

transmitted from a ground station. There is also a limited 'deferred command' capability so that a command sent to the satellite can be stored and implemented much later when the vehicle is out of sight of the ground station. Finally, the spacecraft also contains Sun and horizon sensors so that the position of the spin axis on the sky may be measured and appropriate settings made to the magnetic control coil.

The operations plan involved control and data collection at a ground station at Winkfield in Berkshire. Commands to control the satellite and its experiments are generated at the Ariel VI operations control centre which is located at the SRC's Appleton Laboratory near Slough. They are then sent by means of a data link to the Winkfield ground station. Data from the satellite are received at Winkfield during the four or five orbits each day, out of a total of fifteen, during which the satellite passes over the station. In fact, for reasons that will be discussed later, it has proved necessary to employ several other ground stations in addition to Winkfield in order to control the satellite and to acquire the data.

Of the three experiments, the principal instrument is the ultra-heavy cosmic ray detector provided by Bristol University. Heavy cosmic ray particles pass through a large sphere which contains a specially selected gas mixture and an inner sphere made of plastic. The particles give rise to different kinds of light signals in the gas and in the plastic. The different light pulses are detected by photomultiplier tubes. A comparison of the light intensities allows the charge of each particle to be determined. During the two year mission, a large number of particles will be detected and the resulting charge spectrum, particularly for the ultra heavy nuclei (e.g. uranium) will help to clarify the origin of the cosmic radiation.

In addition to the cosmic ray instrument, the satellite contains two X-ray astronomy experiments. One of these operates in the energy range 0.1 to 1.5 keV and is provided jointly by the Mullard Space Science Laboratory of University College, London, and Birmingham University. It employs four small grazing incidence X-ray reflectors which collect the rays and focus them into four thin windowed portional counter detectors. All of the telescopes view in parallel along the spin axis. The size of the telescope fields of view can be controlled from the ground by selecting different diameter apertures in the focal planes. The instrument is designed to obtain the spectra of X-ray source at low energies and to study their variability.

The Leicester University experiment consists of four sealed beryllium windowed proportional counters which respond to X-rays in the energy range 1.5 to 50 keV. These detectors have fixed fields of view of around 3 degrees determined by fixed collimators. These collimators are also aligned with their axes parallel to the spacecraft spin axis. This experiment has also been designed to measure source spectra and variability at somewhat higher X-ray energies. Its buffer store enables it to make measurements of X-ray variability with very high time resolution.

Both experiments may be operated together to make observations over almost two decades of X-ray energy. However, there are some sources (hot stars, supernova remnants, diffuse interstellar gas) that are more appropriately studied at lower energies.

Early Scientific Results from Ariel VI

The satellite was successfully launched on 2 June 1979 from the NASA range at Wallops Island, Virginia. The launch vehicle was a four-stage Scout rocket and the satellite was injected into a slightly elliptical orbit of 505 km by 550 km which was within specification. The orbital period was 98.5 minutes.

Although the launching was completely successful, the satellite systems have experienced some difficulty which has persisted and which has lead to a reduction in the amount of data being acquired from the X-ray experiments.

Shortly after launch, it was found that certain commands were being spuriously set in the satellite. Unfortunately, this led to the X-ray experiment high voltage supplies being turned off at least once on every orbit. By determining the location of the satellite relative to the Earth's surface at the occurrence time of each spurious turn-off, it was established that the interference was due to transmissions from the Earth. However, the satellite is not receiving correctly coded commands. Instead, it appears that the transmissions are at such a high power level that they are interfering directly with the command function circuits and are not processed in the normal way by the spacecraft command decoders. High power military radar systems are possible sources of this type of interference.

In spite of the problem referred to above, all the experiments are working well and returning useful scientific data. Due to occasional interference with the operation of the tape recorders, the Bristol University cosmic ray experiment is yielding about 80 per cent of the data expected. However, the measurements that have been made are of considerable importance. A preliminary cosmic ray charge spectrum compiled from a thirty-day observation is shown in Figure 5. Here the number of particles detected is plotted against nuclear charge. Some of the elements that correspond to various charges are also indicated. Since the data are taken from a high priority store of limited capacity, the spectrum is accurate for charges above Z = 36. More complete data will be used later to plot the entire charge spectrum. However, a number of interesting features can already be discerned. Several small peaks are apparent. In particular, the one at Z = 78 which corresponds to platinum nuclei can be seen clearly. However, for the very heaviest nuclei, fewer particles than expected have so far been detected.

For many years, astrophysicists have speculated about the nuclear processes that lead to a creation of heavy elements in the Universe. On the one hand, so called 'slow' processes may go on continuously throughout the Galaxy but these would lead to the creation of fewer heavy elements than the 'fast' processes which occur only during violently explosive events such as supernovæ. By establishing the cosmic ray charge spectrum, it should be possible to identify which kind of nuclear process is mainly responsible for the production of the heavy elements. The preliminary results obtained by Ariel VI seem to indicate that both processes are at work and are almost equally important. However, this question should be settled by the analysis of the data from the first year of operation of the Bristol experiment.

While the spurious effects in the spacecraft's command system have reduced the X-ray instruments' data collection to

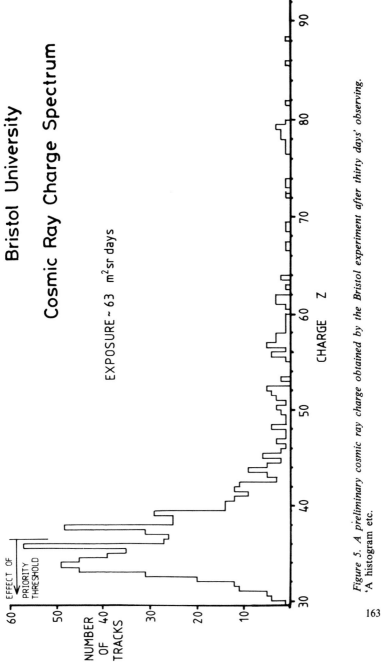

Figure 5. A preliminary cosmic ray charge obtained by the Bristol experiment after thirty days' observing. 'A histogram etc.

around 40 per cent of that expected, both experiments are nevertheless achieving some interesting results. The Leicester University experiment responds to X-rays in the energy range 1.5-50.0 keV. It was designed principally for the study of variable X-ray sources with high time resolution. In addition, since both X-ray instruments view space in the anti-solar direction, collaboration with optical astronomers is highly effective.

Several important results have already been obtained after only six months' operation in orbit. During July 1979, the Leicester group participated in a world-wide 'burst watch' campaign in association with a number of optical and radio observers. An example of an X-ray burst is shown in Figure 6. A burst source, 4U 1735-444, was one of the first objects studied in the campaign. The X-ray signal is notable for its rapid rise and somewhat slower fall. Models of burst production envisage either disturbances in the magnetic field of a neutron star which allows a sudden increase in the rate of accretion or the nuclear detonation of material that has piled up on the neutron star's surface. It is hoped that joint optical and X-ray observations will allow us to discriminate between these views of burst sources.

The Leicester instrument has also observed the periodic X-ray source GX1+4. The variation of X-ray flux is illustrated in Figure 7 in which two X-ray pulsation cycles are plotted against time. The measured period of the source on 9 July 1979 was 112.076 ± 0.01 seconds. This value is compared with previous measurements of the period in 1971 by a balloon-borne instrument and in 1975 by an experiment on the Massachusetts Institute of Technology's SAS-3 satellite (see Figure 8). The period is clearly shortening at a rate of almost three seconds per year. This indicates that the X-ray source is almost certainly a rotating neutron star which is paired with a normal star in a binary system. The X-rays are generated by material which is ejected from the normal star and falls through the intense gravitational field of the neutron star to the stellar surface. The infalling material acquires so much gravitational energy that it is heated to temperatures approaching 10^9K and

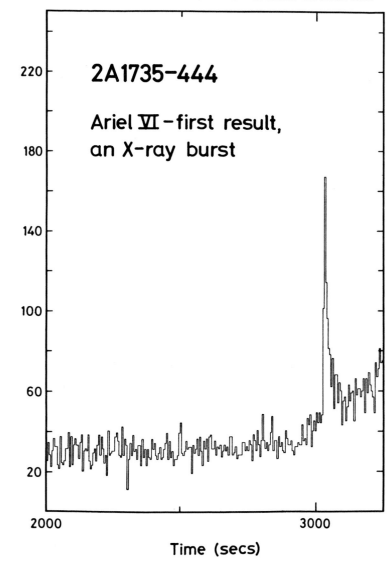

Figure 6. An X-ray burst observed by the Leicester X-ray instrument.

Leicester University Observation of GX1+4

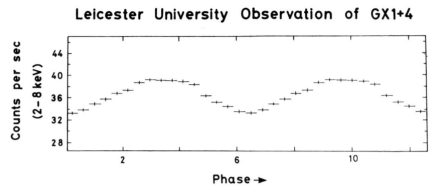

Figure 7. X-ray pulsations from the slowly rotating neutron star source GX1+4 observed by the Leicester experiment.

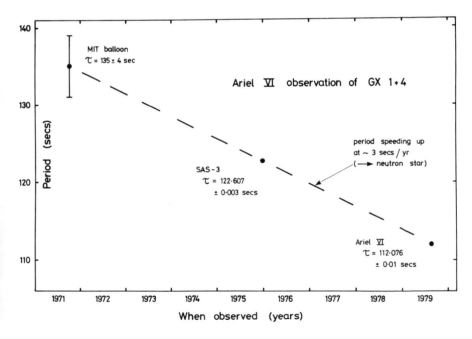

Figure 8. A comparison of the period of GX1+4 measured by Ariel VI with earlier determinations. The rate of change of period with time clearly shows that the pulsating source is a slowly rotating neutron star.

so emits X-rays. The neutron star acquires angular momentum from the infalling material and is therefore gradually increasing its rotation rate. The rate of increase is now well determined by comparing the Ariel VI and SAS III points in Figure 8 and so the X-ray source is confirmed as a rotating neutron star.

Many other studies are in progress with this instrument including observations of sources in the galactic bulge, of active galaxies such as quasars and Seyferts and of clusters of galaxies. In addition, the bright and well-known X-ray source associated with the Crab Nebula supernova remnant has been observed and used to check the calibration of the experiment.

The low-energy X-ray experiment, provided jointly by the Mullard Space Science Laboratory and the University of Birmingham is also only acquiring about 40 per cent of the data expected due to interference with the experiment high-voltage command. Sensitive in the energy range 0.1 to 1.5 keV, the experiment is mainly intended for the study of soft X-rays from high-temperature gas. Such gas is found associated with many sources such as supernova remnants and white dwarf stars. In addition, the instrument also benefits from a high time resolution data acquisition system so that together with the Leicester experiment, it can study variability and source spectra over two and a half decades of X-ray photon energy.

It is well known that at low X-ray energies (i.e. less than 1.5 keV) there is a general or diffuse emission that can be observed from our galaxy. This emission arises from million degree ionized gas that is distributed throughout the interstellar medium. Many individual features are already apparent in this radiation. Some are associated with large radio loops, others with associations of early type stars but, in general, the origin of this hot gas and the mechanisms by which it was heated remain obscure. One of the aims of the low-energy X-ray experiment is to establish the nature and origin of this gas by measuring the spectra and intensity of the X-ray emitting regions throughout the galaxy. A start has been made on these observations and a typical X-ray spectrum is shown in Figure 9. During the course of the satellite's two-year mission, a large

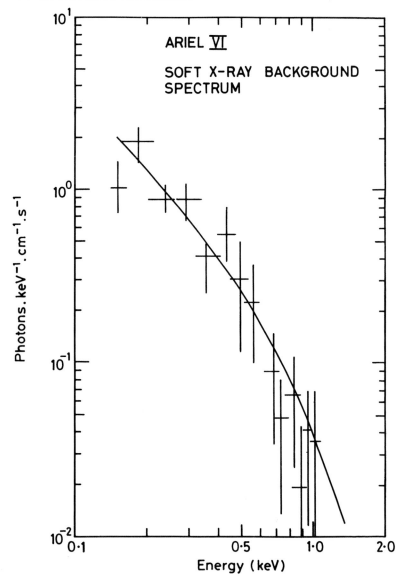

Figure 9. A low-energy spectrum of the soft X-ray background obtained by the MSSL/Birmingham experiment.

part of the Galaxy will be mapped in soft X-rays and we will attempt to relate features in these maps to large-scale optical and radio features.

A number of individual soft X-ray sources have already been observed. One of these is a bright variable source in Cygnus known as Cygnus X-2. This object has been studied previously at energies above 2 keV. A spectrum for the range 0.1 to 1.5 keV is shown in Figure 10. The solid line represents the best fitting single spectrum, a traditional method of describing X-ray data. However, a two component description of the radiation is also acceptable and this is shown in Figure 11a. The emission below 1 keV may be due to radiation from the surface of a white dwarf with a characteristic temperature of around 10^6 K. The instrument is also detecting the tail of a higher energy component due to radiation from shock heated gas at the much higher temperature of 5.10^8 K. An outline model to explain the observations is indicated in Figure 11b. It

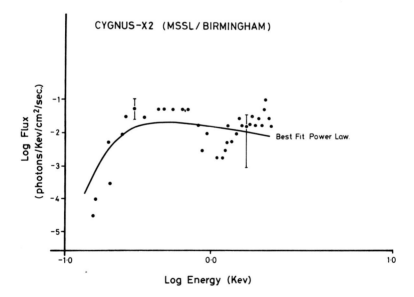

Figure 10. The X-ray spectrum of Cygnus X-2 obtained by the MSSL/Birmingham instrument.

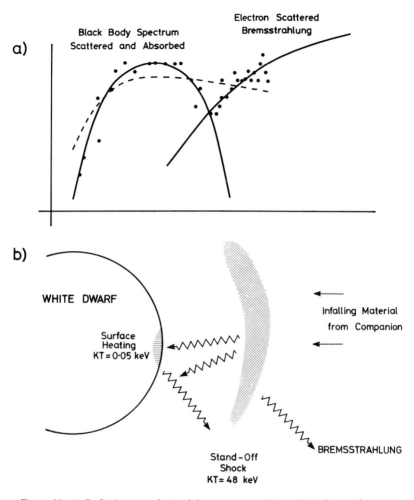

Figure 11. a) Preliminary analysis of the spectrum in Figure 10 indicates that two components of X-ray emission are involved. b) A model of Cygnus X-2 which includes a large companion star and a white dwarf.

is here assumed that the X-rays are generated as a result of material falling from a large companion star on to the surface of a white dwarf. Shock waves are set up due to collisions in the infalling gas and the temperature is thus raised to a very high value with the consequent production of X-radiation by the Bremsstrahlung process. Some of this radiation falls on the surface of the white dwarf and raises its temperature to around 10^6 K thus giving rise to the lower energy component of the observed X-radiation. Observation of these two components helps to demonstrate the presence of a white dwarf in the Cygnus X-2 system.

Many other soft X-ray sources have been observed by the experiment. In particular, the Crab Nebula has also been used as a standard source for this experiment to confirm its calibration. Joint observations (with the Leicester instrument) have been undertaken for a number of Seyfert galaxies and the future programme includes further studies of this kind in addition to observations of purely low-energy sources.

Conclusions

Following a successful satellite launching in June 1979, all three experiments on board Ariel VI are working satisfactorily and, in spite of interference with the spacecraft command, are returning useful scientific results.

The Bristol cosmic ray instrument is acquiring data on the number of ultra-heavy cosmic ray particles that bombard the Earth's atmosphere. During the two-year mission lifetime, this experiment will obtain the cosmic ray spectrum from below iron to beyond uranium for the first time. This information should play an important part in establishing how the heavier elements are created.

The two X-ray experiments provided by Leicester University and by the Mullard Space Science Laboratory of University College, London, and Birmingham University are together covering the X-ray energy range from 0.1 to 50 keV. They are thus capable of studying the X-ray spectra and variability of a wide range of sources. Since both instruments point in the anti-solar direction, collaborative observations with ground based optical astronomers can be effectively undertaken. Having weathered some early difficulties in its operation, the satellite is now embarked on what we hope will be a useful and productive two-year mission.

The Sizes of the Minor Planets

GORDON TAYLOR

A glance at some of the popular books on astronomy may well give the reader the impression that the sizes of the minor planets are well determined. Some diameters, for example, appear to be given to the nearest kilometre. Unfortunately, the truth is very far removed from this happy state of affairs. Determinations of the diameter of Pallas, for example, have varied widely. Whereas in 1894, Barnard found a value of 490 kilometres, Dollfus recently (1971) derived a value of 920 kilometres, almost 90 per cent larger. Yet we now know, from accurate measurements made in 1978, that Barnard's value was much nearer the truth.

The real difficulty in determining the sizes of these bodies is due to the fact that they are relatively small, and that none of the larger ones ever come close to the Earth. In other words, their angular sizes are very small. Only large telescopes at good observing sites can show them as disks rather than points of light. Ceres is believed to be the largest minor planet, with a diameter of about 1,000 kilometres. Yet it seldom gets within 1.6 astronomical units (A.U.) (=240 million km) of the Earth so that its maximum angular diameter is

$$\frac{1{,}000}{6{,}378} \times \frac{8.794}{1.6} = 0.9 \text{ arc seconds}$$

where 8.794 arc seconds is the Earth's angular radius at a distance of 1 A.U. (the solar parallax) and 6,378 km is its linear radius.

There are four main methods of determining sizes, and the first three of these will be reviewed only briefly, since it is the fourth which has provided, though only in a few cases, values which are accepted as of considerably greater accuracy than those obtained by the other methods, and which are the main subject of this article.

Until the middle of this century the only method to give any results was one which employed a visual examination of the minor planet through a telescope which carried some form of micrometer. Barnard used the filar micrometer in 1894-5 to measure the diameters of Ceres, Pallas, Juno and Vesta. In its simplest form, the micrometer consisted of two parallel wires whose separation could be altered by an observer. The measured separation of the wires in linear measure was translated into angular measure so that the angular diameter of the minor planet was obtained. Combining this with the known distance of the body enabled its linear diameter to be determined. However, there are a number of difficulties in the observation. Not only is the angular size very small indeed, but there is also the problem of irradiation. A good illustration of the latter problem occurred during the occultation of Regulus by Venus in 1959. Regulus disappeared, as seen to a large number of observers, as it passed behind the atmosphere of Venus, well above its cloud layer. Yet the measured 'occultation diameter' of 12,338 km led to an angular diameter almost 1 arc second *less* than that tabulated in astronomical almanacs. Hamy, in 1899, used an interferometer to measure Vesta's diameter, while more recently a double image micrometer and also a diskmeter (which uses an artificial image) have been used.

For the past two decades, it has been possible to detect thermal emission from bodies in the Solar System, including some minor planets. Observations of such bodies have been made on infrared wavelengths of around 10-20 μm and the absolute flux from those minor planets measured to an accuracy of the order of 10 per cent. These measurements can be converted to give an infrared diameter if it can be assumed that the body is a smooth, spherical blackbody at the observing wavelength, and that it does not rotate. Departures from these assumptions mean that the infrared diameter could be several tens of per cent different from the true diameter.

Studies of the polarization of the Sun's light reflected from the minor planets has led to the determination of their geometric albedoes, following the formulation of a relationship between the albedo and the shape of the polarization curve at different phase angles. Diameters can then be calculated from these albedoes and the observed magnitudes of the minor planets.

It should be noted that both the radiometric (infrared) and the polarization methods depend on certain assumptions and often give discordant results. For example, in the same volume of the *Astronomical Journal* quite different values are derived from the two methods as will be seen from the table below. The last line of the table gives the value from a stellar occultation.

Table 1

Diameters (km)

Method	Minor Planets			
	Pallas	Hebe	Melpomene	Herculina
Polarimetry	573	195	123	122
Visual and Infrared Photometry	{778 / 730	—	180	284
Occultation	538	195	131	220

Although this sample is too small to be statistically significant, it does indicate the possibility that infrared diameters are rather too large.

The use of photoelectric observations of a lunar occultation of a minor planet to measure the latter's size has been considered for some time but practical problems of observation have so far prevented any useful results being obtained.

If an occultation of a star by a minor planet can be timed from several places on the Earth, then the size of the minor planet can be determined to a very high degree of accuracy. For example, one observer may measure the duration of the occultation to an accuracy of 0.1 seconds. It is then a simple calculation to determine the length of the chord across the minor planet to an accuracy of 1-5 km, depending on how rapidly the minor planet appears to be moving across the line of sight. This is the attraction of using the occultation technique. What a pity that such occultations occur so rarely!

As soon as high precision ephemerides of the minor planets become available (in 1950) it was realized that it was theoretically possible to calculate occultations of stars by these minor planets. For many years after 1950, ephemerides of Ceres, Pallas, Juno and Vesta only were printed to a precision of 0.1 arc seconds and these ephemerides were used in conjunction with star catalogues to predict appulses (close approaches), and actual occultations. Unfortunately, although the ephemerides were of high precision, they were not very accurate and on one occasion in 1954 Pallas was found to be several arc seconds away from its predicted position, so that the track of an occultation originally predicted to cross North America was actually displaced well out into the Pacific Ocean. From all the laborious prediction work done by hand between 1952-67 only one observation was obtained. This enabled a minimum diameter of Juno to be derived. In the 1960s, an observation of Pallas occulting a star made at one station only, again led to a minimum value for the diameter of the minor planet.

The use of computers in astronomy rapidly led to improvements in orbit computations, and also increased the capability of producing predictions. By 1973, accurate orbits of about 30 minor planets were available, mostly computed by the Institute of Theoretical Astronomy in Leningrad, and supplied to the Royal Greenwich Observatory at Herstmonceux, where the prediction work was already computerized.

The first success was obtained in 1977 when Hebe occulted the bright star γ Ceti as seen by several visual observers in Mexico. As a result, the diameter of Hebe was found to be 195 km.

However, 1978 was certainly the year of the minor planet as far as the occultation work was concerned, three occultations by minor planets being successfully observed.

Back in 1977, plans were made to observe the occultation of

a star by Pallas, which was expected to be visible from the U.S.A. on 29 May 1978. The Kuiper Airborne Observatory (KAO) had already successfully recorded two previous occultations, one by Mars in 1976 when the 'central spike' in the light curve was detected and one by Uranus in 1977 when the rings were discovered. Could we predict the track of the occultation by Pallas, probably only 500 to 600 km wide, accurately enough to place the KAO inside the shadow? Now this means that 250 to 300 km could make all the difference between a central occultation and no occultation at all: at the distance of Pallas 250 km is equivalent, in angular measure, to only 0.15 arc seconds. In other words, the position of Pallas relative to the star must be predicted, at least a day or two in advance, to considerably better than this figure to ensure success. The plan was to take plates at Herstmonceux, Flagstaff, and in Germany (hopefully not all three would be clouded out) a few days before the event, when Pallas had approached close enough to the star for both objects to be measured with respect to the same reference stars. This was vital, since even with modern astrometric techniques it is not possible to tie in one area of sky to another area, say 10 or more degrees away, to the accuracy required for this prediction. Although cloudy in Germany, both Herstmonceux and Flagstaff obtained plates, and feverish activity on both sides of the Atlantic on 27-28 May produced consistent predictions. In fact, the final predicted track sent from Herstmonceux on 28 May was only 100 km (= 0.06 arc seconds) from the truth. The occultation was successfully observed, not only by the KAO, but by six U.S. observatories on the ground as well. All the observations were photoelectric (the change in magnitude at occultation was too small to be detected visually). Thus Pallas attained the distinction of being the only body in the Solar System to have had as many as seven photoelectrically measured chords across it using the same star at the same event. From previous work it was realized that Pallas' polar axis was pointing nearly towards the Earth (the angle being about 20 degrees) so that the chords were all measured across the equatorial regions of Pallas. Analysis showed that the equator was elliptical in shape (i.e.

Pallas is triaxial) and that the mean diameter is 538 km.

Only ten days after the Pallas event the minor planet 532 Herculina occulted a star as seen from three places in the western U.S.A. One photoelectric observation and two visual observations were obtained and a diameter of about 220 km was derived from the analysis. It says a lot for the observing conditions in Arizona that the Flagstaff Observatory obtained a good photoelectric record of the event at an altitude of less than 3 degrees!

About two minutes before the occultation by Herculina itself, the photoelectric trace indicated another occultation of duration 5.1 seconds. At the same time a visual observer several hundred kilometres away noticed a series of half a dozen fadings of the star, the most noticeable of which lasted for 4.0 seconds. Using this last observation in conjunction with the secondary event recorded at Lowell and assuming a direction and rate of motion identical with Herculina itself leads to a startling result. The four timed observations (disappearance and reappearance at two stations), reduced to linear measures at the distance of Herculina, fit the circumference of a circle with a very small margin of error. The diameter of this circle is determined at 45.5 ± 2.0 km. It is tempting to suggest that the explanation of these observations is that they are due to a satellite of Herculina. Calculations show that Herculina's gravitational attraction is adequate to retain a body in orbit at much greater distances than that of the projected distance of the observed 'satellite'. It is perhaps a fair summary of current opinion amongst those working in this field to say that the evidence is strong, but not conclusive.

If Herculina has a satellite, then we must realize that we cannot obtain any orbital information from the occultation since we are 'seeing' the two bodies at only one instant of time. However, the photoelectric records do indicate the relative brightnesses of the star and satellite so that as we know the star's magnitude we can calculate the magnitude of the satellite.

The figure shows the circumstances of the occultation, the dots at the ends of the chords corresponding to the observed

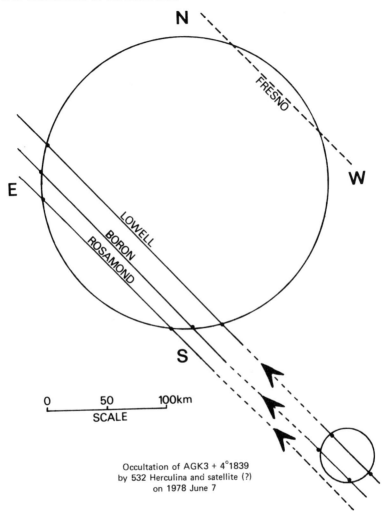

Occultation of AGK3 + 4°1839
by 532 Herculina and satellite (?)
on 1978 June 7

times. Positive observations of the occultation by Herculina were made from three sites, while several observers recorded no occultation at Fresno. Thus the circle fitted to the other observations must be changed into a slight ellipse. The smaller circle has been fitted to the observations of the secondary event. It shows why the observer at Rosamond would not have seen this event. The distance separating the centres of the two circles on the fundamental plane is just under 1,000 km.

The third success came on 11 December 1978, when 18 Melpomene occulted a star as seen from north-east U.S.A. Although Melpomene turned out to have a diameter less than a quarter of that of Pallas we were fortunate in having clear skies over the area and three photoelectric stations inside the track, while the fact that the star turned out to be double meant that we had double the number of chords measured across Melpomene that we might otherwise have expected. From the observations it is possible to derive the magnitudes of the two components, $8^{m\cdot}6$ and $10^{m\cdot}1$, and also the separation 0.04 arc seconds. Photoelectric observations of the occultations of both stars and visual observations of that of the brighter star are quite consistent and, assuming a circular cross-section, the diameter of Melpomene is found to be 131 ± 1 km. There is no doubt that the technique has been tested and proved and we can look forward to other successful events in the next decade!

Jupiter: What the Voyagers Saw

GARRY E. HUNT

Introduction

One of the greatest scientific adventures of all time took place in 1979 when the two Voyager spacecraft encountered Jupiter, sending back detailed information and spectacular pictures of the planet, its satellites and the surrounding environment. The results from this mission were amazing. Suddenly, four tiny point sources of light, the Galilean satellites, became worlds of their own. We found them to be unexpectedly bizarre worlds, each one with its own mysteries that have formed the centre of scientific discussions ever since. Perhaps the most dramatic discovery is the evidence of active volcanism in Io; but this is then closely followed by the unexpected observations of Jupiter's rings.

While observations were being made of these tiny bodies around the planet, vast quantities of information were being obtained about Jupiter itself. Measurements were made for Voyager 1 continuously from the beginning of January until mid-April, and for Voyager 2 the surveillance of the planet then continued until August. Now we have a much better picture of the Jovian weather systems, since time lapse studies have been performed to investigate the growth, decay and movement of the atmospheric cloud systems.

There is no doubt that this has been a very successful mission. For me personally, there is a great deal of satisfaction too, since I have been an experimenter with the Voyager Imaging Team since 1972. The seven years of detailed planning have proved to have been worth while, and, in themselves, serve as a tribute to the enormous efforts of the other scientists and engineers associated with the Voyager project.

In this article, I shall review the emerging picture of Jupiter, its satellites and the surrounding region from the Voyager encounters. As you will see, a great deal has been revealed by these missions.

The Magnetosphere

The magnetic field, with its entrained plasma, makes up the Jovian magnetosphere, huge and variable in size, lying beyond the complex ionosphere. It can extend out to as much as 100 R_J. * Indeed if it could be observed with the naked eye, the Jovian magnetosphere would appear as large as the Sun. Its geometry is, however, similar to the Earth's. The solar wind pressure diminishes as the square of the distance, and this contributes significantly to the size of the Jovian magnetosphere as the solar wind is 25 times weaker at this part of the Solar System. Relatively minor changes in the solar wind pressure can cause large variations in the position of the magnetopause. Pioneers 10/11 first reported crossing this region which marks the entry to the magnetosphere at 100 R_J and then crossing it again later as close as 50 R_J. The Voyager spacecraft found similar variations. For the Earth's magnetosphere to shrink or expand by a factor of two is exceedingly rare, and would be expected only during the most intense magnetic storms.

Table 1

| Voyager 1 | Closest approach to Jupiter | 348,890 km. |
| Voyager 2 | Closest approach to Jupiter | 721,670 km. |

For Jupiter, the electromagnetic environment has a further distinguishing feature, namely that the four Galilean satellites, Amalthea, the tiny 14th and 15th moons and the rings, all reside in this hostile environment. This is quite distinct from the Earth, where our Moon resides in the magnetotail. As a consequence, the interaction of these charged particles with the satellite produces surface characteristics that are unique to the Jovian environment. Charged particles, electrons and protons,

*R_J = 66,770 km (Radius of Jupiter).

trapped in this region that co-rotates with Jupiter, speed around the planet at tremendous velocities. This plasma reaches its highest concentration along the magnetic equator and, in particular, within the torus around the orbit of Io.

Jupiter and Io are connected by a flux tube which carries a current of about 5 million amps. Somehow, Voyager 1 missed this intersection by 7,000 km. Adding further risk to the spacecraft were cosmic rays and other particles originating from Jupiter; and the solar wind manages to leak into the magnetosphere to complicate the situation. Until the 1940s, it was believed that most cosmic rays came from outside the Solar System. Today we believe many are spun off from Jupiter.

The Sun's outer surface is continuously blowing away and sending out streams of hot ionized gases known as the solar wind. When it collides with the outer region of the magnetosphere, the solar wind is travelling at 1.5 million km per hour. Then, suddenly, the particles are slowed down to a mere 400,000 km per hour as the particles are deflected.

Before Voyager, it was thought that some particles would leak through the tail end of this shield and be carried right into Jupiter's polar regions, creating the type of auroral displays familiar to us on Earth. While Voyager did indeed find auroral displays, they are triggered not by the precipitating polar wind particles, but by electrons streaming in from Io's torus. This situation, coupled with the additional observation of extensive radio spectral arcs (~1 to 30 M Hz) occurring in patterns correlated with Jovian longitude, together with kilometric radio emissions generated by plasma oscillations in the Io plasma torus, indicates that the magnetosphere is a source of radio emission, and a sink for Ionian material.

One significant change in the radiation environment was that at the time of Voyager 2, it was three times stronger than at the time of the first spacecraft encounter, just four months earlier. Remember also that Voyager 2 was twice as far from the planet as its sister spacecraft. There is little doubt that had Voyager 1 experienced this enhanced level of radiation, it would have created serious problems, even failure. With the Pioneer spacecraft, we have now successfully run the Jupiter gauntlet four times; let us hope we can continue this record in the future.

The Atmosphere

Jupiter is a fluid planet, composed mainly of hydrogen and helium in (near) solar proportions with small but important amounts of strange exotic constituents such as ammonia, methane, water vapour, ethane, acetylene, phosphine, germanium tetrahydride and carbon monoxide. All these constituents were detected originally from spectroscopic studies. Now, with the Voyager spacecraft, we have our first opportunity to examine the spatial distribution of the constituents. This information may ultimately prove valuable in our understanding of the Jovian cloud colours, since most of these constituents have been found in and around the cloud tops.

The infrared radiometer infrared spectrometer (IRIS) on the Voyager spacecraft has found a significant variation in the ethane and acetylene distributions. Between the measurements on the individual spacecraft, we found that the ethane to acetylene ratio increased between the encounters, and was also significantly larger at the polar regions than at lower latitudes.

Haze layers are also evident in the UV maps made at 2,400 Å. The polar regions are surprisingly dark, suggesting that absorbent material must penetrate above the 40 mb level. Imaging studies had certainly suggested that these hazes probably extended into the mesosphere where the local pressure was ~3 mb. These layers of absorbent material may be created by photochemical processes and, in addition, they could possibly result from material originating in the Jovian ring.

Thermal maps of the planet at 150 mb and 800 mb (cloud tops) show both local and large-scale structure which appears to correlate with the visible appearance of the planet. Well-defined features, such as the Great Red Spot, are evident as colder regions, with temperatures of ~144 K, some 2 K below the neighbouring areas. However, the gross belt/zone structure observed in the visual images is apparent in these maps. Relatively cold regions (~141 K) are found at 20-35 degrees N and 20-35 degrees S corresponding to bright zones in the images (see Figure 1). The warmest temperatures ($\sim$149 K) in both maps occur in a region between approximately 15 degrees N and 15 degrees S, which is associated with a visually dark equatorial belt. These temperature contrasts may also relate to the dynamical heating and cooling of the atmosphere.

Dynamical changes in the atmosphere are also evident in the 5 μm maps of the planet, which probe the structure of the clouds, since they correspond to a region of the spectrum where there is negligible atmospheric absorption. The deepest regions of the atmosphere that we can probe correspond to emission temperatures of about 258 K and occur in the north equatorial belt. The coldest areas are the Great Red Spot and the white zones which appear featureless at this wavelength. Since we are probing the atmosphere, sudden changes in the 5 μm maps are found on time scales of just a few planetary rotations. However, the NEB has remained a region of high 5 μm emission throughout the Voyager encounters.

Without doubt, the largest unsolved problem of the Jovian atmospheric system must be understanding the origin of the colours. The discovery of lightning storms occurring on a global scale may be important in this context. We believe that the action of UV radiation and lightning may be the important energy source that energizes the photochemical reactions involving ammonia, methane and some sulphur-bearing compounds. It has been thought that hydrogen polysulphide H_xS_y or ammonium polysulphide $(NH_4)_xS_y$ are generally yellow, orange or brown according to the local temperature and could be involved in these complex chemical reactions. The key to this cycle is the identification of a sulphur compound, of which H_2S is a likely candidate. We are still carefully searching through our data.

Meteorology

The weather systems on Jupiter appear to be very different from those found on the Earth and the terrestrial planets, and a completely new perspective has been obtained from the incredible images obtained during the Voyager encounters in 1979. With the nearly continuous monitoring of atmospheric pheno-

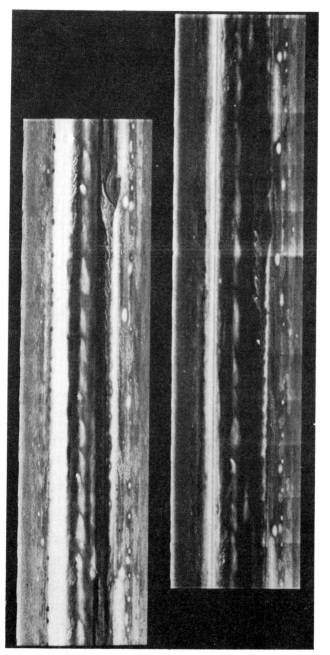

Figure 1. Cylindrical projections of Jupiter representing Voyager 1 (1 February 1979) and Voyager 2 (28 May 1979). Each image covers 400 degrees of longitude with the 0 degree position at the right-hand edge.

mena for almost eight months, a detailed picture of the horizontal atmospheric motions was obtained, down to scales of just 3 km. These observations have coupled together space and time, so that now the complete four-dimensional picture of the planet's motions has been obtained, so that we are now in a position to unravel the complicated meteorology.

Jupiter emits nearly twice the energy it receives from the Sun. This additional energy has an important rôle in the meteorology, since it means that the planet's weather is affected both by an internal heat source and external solar radiation. Also, effects of the planet's rotation are very strong, since the period of rotation is approximately 10 hours. Consequently, on a large scale, the motions are predominantly zonal, but as can be seen from Figure 1, there are regions of significant meridional motion. On a global scale, there is little, if any, pole to equator energy transfer at the visible cloud levels, and this is a major difference between the Earth and Jupiter. Indeed, the Pioneer 11 measurements showed that there was a very small difference between the effective temperature of the planet at the equator and poles of no more than 3 K. We also see the gradual breakdown of the dominant symmetric belt/zone pattern beyond ±45 degrees latitude, where the internal heating becomes the dominant driving mechanism.

The temperature contrasts between the belts and zones are also small. However, the location of the maximum contrast does vary, and between the Pioneer 10/11 flybys of 1973/74 and the Voyager flybys of 1979, it has shifted hemispheres. Previously, the largest contrast, which was about 3 K, occurred between the south equatorial belt and south tropical zone, while now it is at their northern hemispheric equivalent latitudes. These horizontal temperature contrasts, which may be related to the release of latent heat of condensation, probably play an important rôle in the planet's weather system, creating one of the most surprising results of all; namely, that the weather systems on Jupiter and the Earth may be driven by similar processes.

From the analysis of the Voyager data, we have found that energy is being transported from the small-scale systems into the mean zonal flow. Jupiter is doing locally the same as happens on the Earth on a global scale. Accurate studies of the jets have shown that their velocities, measured from small-scale cloud features only ~100 km across, yield the same values as the Earth-based observers have found over many decades tracking features many thousands of kilometres across. Consequently, this means that we are actually measuring the mass motions of the atmosphere. The stability of the jets over many decades is also at variance with the visible appearance of the planet, which seems to change in just a few rotations. For this situation to occur means that the jets are controlled by deep-seated forcings several hundreds of kilometres below the visible surface.

During the Voyager ·flybys, the highest velocities were measured in the stable westerly jets, particularly in the north temperate region, with values of ~150 ms⁻¹. Although the equatorial jet moves at a speed of ~100 ms⁻¹ relative to the surrounding regions, the flow is not symmetric and clouds at the Equator actually move some 10-20 ms⁻¹ slower. The train of dark spots seems to roll along the northern edge of the N.Tr.Z. at about 20 ms⁻¹, while the vortices that were seen to approach the Great Red Spot at the time of the Voyager 1 encounter at 50-60 ms⁻¹ before being accelerated to 100 ms⁻¹, were squashed and finally destroyed (Figure 2).

The overall picture of Jupiter shows a great variety of cloud morphologies of varying colours, sizes and shapes; and super-imposed on the system are large-scale spots, namely the white ovals and the Great Red Spot. We have found that all the large-scale spots are alike. The Great Red Spot is simply the largest of a family of giant circulation gyres that behave like strong high pressure systems on Earth, but are larger and live longer. These features probably form from the large-scale weather systems that constitute the parallel bands of cloud, the belts and zones. Why do Jupiter's cloud features last so long, in some cases decades and even centuries, when Earth clouds disappear in a few days? Again it is so cold on Jupiter, where the cloud tops are at about $-120°C$ ($-184°F$) that less energy is lost by these systems. Also, they penetrate deep into the

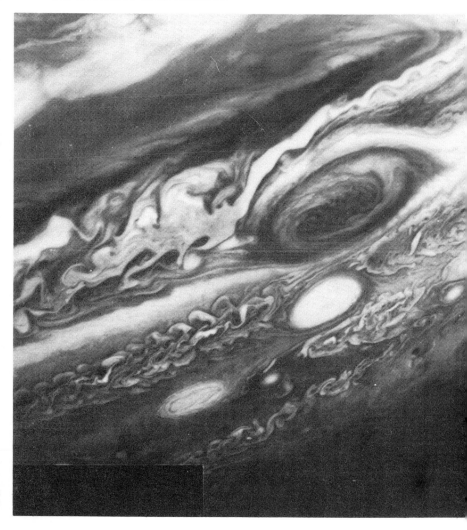

Figure 2. The Voyager 2 image of the region around the Great Red Spot.

Table 2
Physical and Orbital Characteristics of Amalthea and the Galilean Satellites

Satellite	Diameter (km)	Mean Distance From Jupiter	Orbital Period (days)	Density (g/cm³)	Mass (Moon = 1)	Closest approach (km)		Best Resolution (km per Line Pair)	
						Voyager 1	Voyager 2	Voyager 1	Voyager 2
Amalthea	155x270 (±8)	109,900	0.49	?	?	420,100	558,270	7.8	11
Io	3,638 (±10)	350,200	1.77	3.53	1.21	18,640	1,127,920	1	21
Europa	3,126 (±10)	599,500	3.55	3.03	0.66	732,270	204,030	33	4
Ganymede	5,276 (±10)	998,600	7.16	1.93	2.03	112,030	59,530	2	1
Callisto	4,848 (±10)	1,808,600	16.69	1.79	1.45	123,950	212,510	2.3	4

atmosphere, which is completely fluid, so there is no surface to break up the clouds. But they will not last for ever. Even the Great Red Spot, now a mere 21,000 km long, is half the size of a century ago. One day it will disappear.

So we are left with the problem of the colour of the Great Red Spot. For the moment, we can only guess, but a possible explanation must come from the presence of phosphine in the atmosphere. As it is drawn up through the atmosphere from the deep layers, the action of sunlight could ultimately lead to the production of red phosphorus. The absence of a multiple of red spots may simply be because the other features do not penetrate as deeply.

The dynamics of the atmosphere of Jupiter are of interest not only to meteorologists, but to exobiologists and chemists too. The discovery of lightning storms all over the planet means that the chemical precursors to the molecule of life are being made continuously in Jupiter's clouds. It is certainly possible that the warmer regions of the atmosphere, beneath the visible clouds, are similar to the early Earth. We can still fantasize about organic compounds living in the clouds, but the atmosphere may be too unstable to allow the formation of the complex biological molecules. There is still a great deal to learn.

The Satellites

Jupiter is surrounded by an extensive system of satellites, of which four inner members, discovered by Galileo in the 17th century, are planet sized. Their names, moving outward from the planet, are Io, Europa, Ganymede and Callisto. With Jupiter, they form a solar system within a solar system. Before Voyager, these satellites were no more than point sources of light with surface detail impossible to observe, even through the largest telescope on Earth. So it is no wonder that Voyager has shown us that these bodies are the strangest worlds we have ever seen (Figure 3).

Callisto is a giant ice ball, with a totally cratered surface and ancient scars which were produced during the formation of the planetary system. The craters do not seem to have the same

Figure 3. A schematic image of Jupiter and the Galilean satellites, constructed from the Voyager images.

sharp rims and deep floors as we find on Mercury and the Moon. There are also two huge basin structures, about 600 km in diameter, with pressure ridges radiating outward for more than 1,000 km. The central portion of the basin has become filled in with material through the passage of time and cratered by small impacts.

Ganymede has a more varied surface than its ice neighbour Callisto, with fewer craters. Here we find a grooved surface formed as the crust has cooled and contracted leaving an ice wrinkled pattern. There are not many craters visible, since the ice crust has oozed up from below leaving faint ghostly images of the crater impacts. However, there is a large impact basin similar to that found on Callisto which has been partially obscured by subsequent surface melting and movement.

But the real enigma of the set of bodies is Europa. After the first spacecraft encounter, all that could be seen were enormous and mysterious stripes criss-crossing the satellite. As the subsequent pictures were received on the Earth with improving resolution, we found that Europa resembled a cracked white billiard ball. In fact, these markings looked as if they had been painted on, for the satellite seems remarkably flat. Also, there are very few craters, and after considerable study only three potential impact craters have been found. How can Europa maintain such a young face? We can only guess at this stage. But it is likely that Europa has a slushy icy surface that expands and cracks on a grand scale due to the forces created by the neighbouring bodies. The upwelling water or soft ice will then flow through these fissures and quickly erase its craters. A strange and surprising world.

There is no doubt that the anomaly of the Galilean satellites, and of the solar system as a whole, is Io. Io was expected to be heavily cratered, but instead we have not found one such feature, for the surface is extremely young. Instead we have discovered the most active volcanic body in the Solar System, with as many as eight active volcanoes observed at a single time. Sulphurous fountains were seen spurting material at ballistic speeds to altitudes of nearly 300 km. Giant streams of molten sulphurous lava flow over the satellite. Pulled and tugged by the gravity of Jupiter and Europa, the surface of Io seethes with tidal forces which heat the upper layers and induce volcanism. Vast electrical currents which flow from Jupiter to Io carrying 5 million amps of electricity may also contribute to the generation of explosive Ionian plumes. The power contained in this electrical energy is equivalent to about 20 times the combined generating capability of all the nations on Earth. Could Io have always been as active as it is now? Certainly, this would seem strange, since all the bodies we know well, the Earth, Moon, Mercury and Mars, have had variable periods of volcanic activity. Is Io special in some way? Possibly the tidal actions could have continued for more than 4 thousand million years, aided by the electrical heating in a self-limiting system. There is no doubt that Io will be the centre of study for future missions to this region.

What happens to the material ejected so violently and continuously in these plumes? Some of the sulphur and sulphur dioxide falls back on to the surface while the remainder forms a doughnut-shaped visible torus around Jupiter along the satellite's orbit. Here we have found clouds of ionized sulphur, oxygen and hydrogen.

Even closer to the planet, we have seen for the first time the tiny moon, Amalthea. It proved to be a dark red and rocky cigar-shaped body, bearing scars, which suggested that once it may have been part of a larger object.

But there were still even greater surprises in store as we searched closer to the planet. Voyager found that Jupiter is surrounded by a system of rings. This was totally unexpected. They are quite unlike the rings of Saturn. The Jovian system of rings is no more than a kilometre thick, and is composed of dense microscopic particles that extend down to the cloud tops. But where do they come from? Perhaps they could flow in from Io's volcanoes; or material from neighbouring moons; or even débris from a captured comet. But with such an uncertain supply of material, it is indeed possible that the Jovian rings may not always be present. Right at the edge of the rings, a further spectacular discovery has been made. Jupiter's 14th moon is a tiny, potato-shaped object only 30-40 km. in diameter. Its presence is important since it sweeps up the particles at the rings' outer limits. Travelling around Jupiter in only 7 hours 8 minutes, it is the fastest moving object in the Solar System. Jupiter's 15th moon has recently been found between Io and Amalthea, only 70-80 kms in diameter, its period is 16hrs 16min.

Conclusions

For the Voyager spacecraft, the first part of their grand tour through the outer Solar System has been accomplished. The next stop is Saturn, which Voyager 1 reaches in November 1980 before heading out of the Solar System. Voyager 2 will reach Saturn in August 1981, then on to Uranus which it will reach in January 1986, and finally Neptune in September 1989. The possibility of these encounters has been made a reality by the successful flyby of Saturn in September 1979 by the Pioneer spacecraft.

These brilliant interplanetary messengers, the Voyagers, have shown us new worlds and have captured the imagination of us all in their thrilling voyages of discovery. Now the stage is set for a thrilling trip past Saturn. We are lucky to live and work in such an exciting time in history.

Acknowledgements

I wish to thank my many colleagues on the Voyager project, with whom I have worked for the past seven years, and without whom this mission would not have been possible. My research is supported by the Science Research Council.

References

I have listed below the main references to Voyager scientific results which should be consulted in conjunction with this article.

Science, **204,** 913-1008 (1979)
Science, **206,** 925-995 (1979)
Nature, **280,** 725-804 (1979)
Geophys Res. Letts **7,** 1-61 (1980)
J. Geophys. Res., April 1981
The Atlas of Jupiter, G. Hunt and P. Moore (Mitchell Beazley) 1980.

Symbiotic Stars

DAVID A. ALLEN

My dictionary defines *symbiosis* as 'a mutually beneficial partnership between organisms of different kinds'. For me, the word is inextricably associated with memories of California. I remember, for instance, the coast road south from the San Francisco Bay area, which passes the stately home of the Hearst family. Pattie, you will recall, was abducted by a small group of people who dubbed themselves the Symbionese Liberation Army. Too, I remember the nature trails in the various National Parks of California. Each of these trails would, at some point on its route, pause the ardent explorer before a lichen, and the explanatory leaflet would there tell how that lifeform is an example (I sometimes thought *the* example) of symbiosis. In the case of lichen the cohabitation is that of a fungus and an alga.

But more than by these examples, California and symbiosis are, for me, linked by the Mount Wilson Observatory. There, in 1971, I exploited the then novel technique of infrared astronomy to study various unusual stars which had been discovered three decades earlier by Paul Merrill, at the same observatory. Amongst these were the symbiotic stars.

Merrill chose the adjective 'symbiotic' to describe a rather select group of about a dozen stars, most of which he had himself discovered, and which exhibited remarkable properties. These stars appear to be single, yet they display to the spectroscope two discrete and mutually exclusive temperature régimes. At one extreme, there are clear indications of the presence of titanium oxide, an ingredient of many white paints and a material which is decomposed at temperatures much in excess of 3,500°C. At the other extreme, symbiotic stars contain gas which has unquestionably been heated by an energy source at nearly one million degrees. How can these two temperature régimes be combined in a single star?

In addition to their spectroscopic properties, the symbiotic stars found by Paul Merrill were variable in their brightness. Every few years a symbiotic star would brighten by a factor of perhaps thirty or forty. The brightening might take a couple of days, but the star then required several months to return to its normal intensity. During this outburst, the spectroscopic properties would change. At times, there would be evidence neither for very hot gas nor for titanium oxide; instead, a single temperature slightly greater than that of the Sun was indicated. What did this mean?

As always when scientists tackle a perplexing problem, a plethora of ideas and theories has eventuated. Some astronomers have attempted to devise clever explanations of how a single star may indeed display two temperature régimes. These theorists generally took a lead from the Sun. In the Sun, we see a surface whose temperature is not much too high for titanium oxide to survive. The Sun is surrounded by a tenuous atmosphere whose temperature exceeds one million degrees. Because this atmosphere is so very tenuous, we are unaware of it except during a total eclipse, or by using special techniques. All that is required to explain the symbiotic stars, therefore, is a mechanism whereby somewhat cooler stars can equip themselves with a more prominent atmosphere. No really satisfactory way of doing this has been thought of, however.

Other astronomers pursued instead evidence that there really is a pair of stars in the symbiotic systems. To date, only two clear examples have presented themselves, the stars AG Pegasi and AR Pavonis.

AR Pav is irrefutably a pair of stars, for it is an eclipsing variable. As the two stars orbit, they pass in front of one another so that alternately some of the light from one is shut off by the other. A study of AR Pav by John Hutchings and the late David Thackeray showed that one star is large and cool, with a surface temperature of about 3,000°C. The other star is very hot, but is so small that it is not recorded directly, but only by its heating of nearby gas. In the case of AG Peg the same situation has been revealed indirectly, by the careful measurement of the velocities of the various features in its spectrum. Measurements over several orbital periods were required in order to disentangle the details, and since the orbital period of AG Peg is 820 days, this was a long job.

Because of the clear evidence of duplicity in two symbiotic stars, and because, too, of the difficulties encountered in theories which entertain single stars, the majority of astronomers today would favour the double star interpretation. That, however, is not the end of the story. Why, for example, do these stars undertake their variations of both brightness and spectral properties? And why does one component have a temperature of a million degrees when stars that hot are unknown except in the centres of a few large planetary nebulæ? Why, too, have not more symbiotic stars been demonstrated to be binary?

Over the thirty-odd years that symbiotic stars have been known, quite a few more have been found. Indeed, the present total is about 120, of which three are members of the Magellanic Clouds, our neighbouring dwarf galaxies, whilst the remainder are stars within our own Galaxy. The discovery of new examples, perhaps not surprisingly, has added complexity to the picture, and has posed further questions. Most of the newly-discovered examples show no brightness variations—at least not on the time-scales we can witness. Others have undertaken in historic times a single outburst in brightness, in some cases a thousandfold increase, lasting decades. In fact, AG Peg is an example of this type of object. It brightened in the mid-nineteenth century from ninth to sixth magnitude; today AG Peg is eighth magnitude, and is fading slowly. One forms the impression that all symbiotic stars have outbursts, but that in some cases the outburst lasts only a few weeks, whereas in others the outburst is maintained for centuries, or at least long enough so that we see no significant change over several decades. This range of time scales must also be explained.

I have devoted much of the last three years to studying the symbiotic stars. In an attempt to define better their characteristics, I have attacked them (so to speak) from several angles. When I started in research astronomy not much over a decade ago, it was still normal for an astronomer to work only in his restricted field, and to apply that field to a range of different celestial objects. Radio astronomers did nothing but radio astronomy; infrared astronomers made measurements only at infrared wavelengths; and those trained in optical observing considered it *infra dig* to spoil their desk tops with infrared or radio data. Things are different today. More and more of us are finding it preferable to branch out into the various disciplines, and to apply all of them to the problems on which we are working. In trying to piece together the jigsaw of the symbiotic stars I have used the Parkes 210-foot radio telescope, and made a variety of optical and infrared observations on the 158-inch Anglo-Australian telescope. More recently, I have been securing observations with the International Ultraviolet Explorer and the Einstein X-ray satellites. The picture I shall sketch in the remainder of this article seems to me the best interpretation of all these data, though it is by no means the only interpretation which could be concocted, and it may one day be proved wrong.

In their dotage, stars like the Sun pass through a phase in which they are known as red giants. Their girth expands dramatically, and their surface temperature falls to about 3,000° C. This condition is not very stable, and thus short-lived on the astronomical scale of time, mostly because of the antics of the star's core. Deep within the red envelope, a tiny dense core has burnt up all its hydrogen to become helium, is beginning to burn helium to form other elements, and is shrinking. Already the core has squeezed most of the weight of the star into a volume about the size of the Earth. The burning of helium is rather violent, and causes most of the outer layers of the star to be thrown off. The naked remnant core, hot and dense, is a star we call a white dwarf. White dwarfs are quite common in our Galaxy. The age at which the red giant or white dwarf phases come to a star depends upon several factors. It is quite possible for two stars which were formed together to reach these stages at different times, just as twin brothers at age

60 may appear to an outsider to be of quite disparate ages. Thus we can expect, occasionally, to find pairs of stars one of which is a red giant whilst the other has become white dwarf. Symbiotic stars are examples of such a situation. They are further distinguished by the fact that the two stars orbit quite close to one another.

The gas which forms the outermost layers of the red giant finds its loyalties divided, for whilst it rightly belongs to that star's core, the gravitational attraction of the companion white dwarf is always, seductively, present. Sometimes the balance is tipped in favour of the white dwarf. A small stream of gas will peel away from the red giant and spill on to the white dwarf.

What happens when the red giant's hydrogen is poured on to the hydrogen-deficient white dwarf? To answer this, we have to rely on computer simulations. These computations, performed by the Polish-American astronomer B. Paczyński, indicate that, for some while, the hydrogen just accumulates quietly on the white dwarf's surface. Then, quite suddenly, it explosively burns to form helium, exactly as in the hydrogen bomb. The explosion is known as a hydrogen shell flash. When it happens, the white dwarf heats up to about a million degrees. Depending on the rate at which hydrogen has spilled on to the white dwarf, the shell flash may pass in a few days, or it may last for centuries.

The sudden increase in temperature of the white dwarf causes the gas near it, including any that is in the process of spilling over from the red giant, to burst into light, rather like a street lamp when a current through it is switched on. This is the brightening we see in many symbiotic stars. In the initial explosion an opaque, enveloping cloud of gas not much hotter than the Sun may be formed, but this expands rapidly and soon becomes transparent, revealing the hotter gas within. Hence are explained the spectroscopic changes seen during the outburst in many symbiotic stars. So long as gas continues to spill over from the red giant, a continuous series of hydrogen shell flashes will arise. Thus the outbursts repeat, as indeed is the case in many symbiotic stars.

I have yet to explain why so few symbiotic stars show direct evidence of being double systems. In order that the white dwarf is able to suck gas from the red giant, the two stars must lie in close proximity. Their separation is such that no telescope on earth could separate the two. When double stars are too close to be resolved apart, however, the optical spectroscope usually can help. As the stars orbit one another, their velocities to and fro cause spectroscopic changes. Features of their spectra shift in wavelength according to that hallowed principle known as the Doppler effect.

Red giants have rather fuzzy spectra. It is difficult to measure changes in these, and hence to detect any orbital motion of the red giant. Instead, astronomers have concentrated their efforts on the hot gas. Unfortunately, the situation there is not much better. The gas is dashing about, and probably spiralling around the white dwarf as it spills on to it. Its swirling motion is at least as fast as the orbital motion of the white dwarf, and varies according to the exact details of the spillover. These large, erratic motions camouflage the orbital motion, and make the job of measuring the latter almost impossible.

On the other hand, of over 100 symbiotic stars known, we would expect to be viewing at least ten at an aspect where the red star eclipsed the hot gas. It is surprising, therefore, that only AR Pav is known to exhibit eclipses. Clearly there remains some work to be done here, and in addition to professional work, it would be useful for amateurs to monitor some of the brighter symbiotic stars for variability. A recent report (by a Russian astronomer) that eclipses are present in CI Cygni will, if confirmed, add a second example.

It is perhaps instructive to explain why the cause of different astronomical techniques was so relevant to disentangling this rather complex story. At infrared wavelengths the cool stars contribute virtually all of the light of the systems, and can therefore be studied: they appear to be quite normal stars. At the shorter wavelengths accessible to the human eye, the hot gas near the white dwarfs is generally most prominent. Because the amount of gas and the details of its movements and physical conditions are constantly changing, this is the most difficult wavelength region to understand. As we move into the ultraviolet and X-ray domains we can study the white dwarf itself, hence the interest in these satellite measurements.

The radio observations also refer to the gas, but in this case not to the gas near the white dwarf, but rather to much more distant gas which has been thrown off by the red giant and is now illuminated by both stars. To detect the gaseous envelopes of stars at radio frequencies is a very difficult undertaking. Radio astronomers have sought 'radio stars' for many years, mostly without success. It turns out that the symbiotic stars are amongst the easiest of all stellar systems to be detected by radio telescopes: they figure prominently in all lists of radio stars.

The term symbiosis is a good one. Without the red giant to feed it, the white dwarf would be an insignificant object. At the distance of even the closest symbiotic star the white dwarf alone would be too faint for any but the largest telescopes to record. Conversely, without its companion white dwarf, the red giant would seem like just any other red giant—one of 100,000 in our Galaxy. In choosing to call these objects symbiotic, did Paul Merrill already suspect them of duplicity? I cannot say. Merrill, although a prolific writer, does not seem to have left us any thoughts on this matter. He was a careful scientist who committed to paper only those facts he was confident were correct. Yet I think he felt all along that single star interpretations were inappropriate. Why else would Merrill have tried so hard to find binary motion in AG Peg?

There remains to be noted only that this interpretation of the symbiotic stars is tempered very much by modern thinking. Interacting double stars—i.e. those in which gas flows from one star to the other—are in vogue to explain many peculiar celestial phenomena. The subject has become a bandwagon during the last decade or less, and inevitably there is a risk of so popular a concept being applied to inappropriate cases. Amongst the peculiar systems which have been successfully explained in this way are the so-called dwarf novæ, or U Geminorum variables (of which SS Cygni is the brightest and best known), and those systems which radiate gross amounts of energy at X-ray wavelengths. But that is a story for another *Yearbook*.

Cosmic Rays of the Highest Energies

A. W. WOLFENDALE

Early Days
The discovery of cosmic rays is one of the romances of modern science. At the turn of the century there was much interest amongst physicists in the behaviour of gases at very low pressures when subjected to strong electric fields. The observations had already led to the discovery of X-rays and the electron, discoveries which were to have great practical importance as well as academic interest, and it was quite clear that more discoveries were just waiting to be made. Technical improvements allowed the introduction of more and more sensitive electrical instruments, and improved insulators enabled leakage currents to be reduced, but researchers found that no matter how carefully their electroscopes were shielded, there was still the residual current passing through the gas between the electrodes. It was this mundane leakage that was to lead to the discovery of the remarkable 'cosmic' radiation, the study of which still occupies the waking hours of many academics and which continues to give rise to new discoveries.

The actual identification of the radiation causing the breakdown in the electroscopes as due to an extra-terrestrial agency came from the high altitude balloon experiments of Hess in 1912. Hess found that the leakage current in his instrument decreased with height in going from ground level to about 700 metres, and then increased steadily. The initial decrease was identified with the effect of radioactivity from the Earth but the increase with altitude at greater heights was clearly due to some other cause. Hess made the correct diagnosis that the radiation was coming in from above the atmosphere, and because of the absence of any difference between day and night readings he correctly inferred that the radiation was not coming from the Sun. Thus, the cosmic origin was clear.

Even in the early experiments it was realized that the radiation was extremely penetrating, and in so far as gamma rays were the most penetrating radiation known at the time, it was natural to interpret it as some extreme form of gamma radiation. In fact, a number of years were to elapse before it was shown that the so-called 'radiation' was composed not of quanta, but almost entirely of atomic nuclei, largely hydrogen (protons). Rather remarkably, it was not until the late 1960s that a tiny component of gamma rays was discovered in the cosmic radiation (see my article in the 1980 *Yearbook*) and the bulk of the work carried out has been on the predominant particle component.

An exciting development occurred in 1938, when the French physicist Auger discovered showers of particles arriving at the Earth's surface simultaneously. The showers, the so-called 'extensive air showers' or EAS, are produced when a single very energetic particle enters the Earth's atmosphere and initiates a whole cascade of interactions with the atomic nuclei of the atmosphere. These initiating or 'primary' particles can have very high energies indeed; even in the early experiments energies of 10^{16}eV were commonplace (such an energy is about a million times the average energy of individual primaries) and the present record for the highest energy recorded is, as we shall see later, about 10^{20}eV.

Most of what is discussed later is concerned with these very high energy particles—the highest energies known to mankind—but first let us look at some rather general features of cosmic rays.

Cosmic rays—the general scene

The reasons for studying cosmic rays are many and varied. Until the mid-1950s, the high-energy physicist looked here for his new elementary particles and the positron, the muon, pion and the so-called 'strange-particles' were all discovered in the cosmic ray 'beam'. The advent of high-energy accelerators has meant that detailed studies of elementary particles have been made with the machines, but there is still quite significant effort devoted to using cosmic rays to give general guidance as to the properties of high-energy particle interactions at the higher energies available in cosmic rays (currently, accelerator experiments involving protons hitting protons are limited to about 10^{12}eV). Whereas the high-energy studies have contracted a little, the astrophysical aspects of the subject have attracted ever-increasing attention.

The astrophysicist is interested in all the various properties of the primaries: their masses, energies, directions and intensities as a function of time. The interest stems from a desire to know where the particles have come from (supernovæ? pulsars? distant galaxies? quasars? . . .) and how they have propagated through the intervening space. It is salutary to examine the magnitude of the energy brought in by cosmic rays; a simple integration of the energy spectrum (Figure 1) gives about 0.5eV per cm³ for the energy density and this is just about equal to the energy density of visible starlight! When one stands outside on a clear moonless night and, well adapted, scans the sky it is hard to realize that completely invisible particles are bringing in just about the same amount of energy as appears so brightly from the stars. A good question is to ask why, then, does optical astronomy occupy such a major rôle in examining the Universe and the study of cosmic rays occupy the attention of only a favoured few? The answer is clearly not that the cosmic rays are invisible to the eye—the new astronomies (radio, infra-red, ultra-violet...) deal also with invisible quanta, but they, like optical astronomy, are very well patronized. The answer lies in the fact that the magnetic fields in the Solar System and in the Galaxy in general so deflect the cosmic ray particles that their arrival directions bear hardly any relationship to the directions of the sources. Thus, an analysis of where the particles come from is a very uphill task indeed. Having said that, the fact remains that there is a lot of energy in cosmic rays and this must be taken into account when analysing the energy balance in the Galaxy (and in the Universe in general) and one can still make some progress in the search for

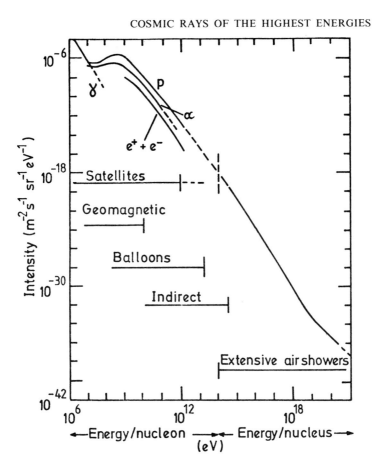

Figure 1. The cosmic ray spectrum with an indication of the techniques used in their study. Protons predominate at low energies (below ~ 10¹²eV) where direct mass determination is possible. Above 10¹⁵eV the composition is very uncertain but it is probable that protons are still the main component.
The spectra of α-particles, electrons (e⁻), positrons (e⁺) and γ-rays are also shown.

sources, particularly at the very highest energies where the effect of the magnetic fields on the particle trajectories is smallest. More about this shortly.

The physicists in other areas also have interest in cosmic rays. For example, cosmic rays interacting in the atmosphere produce secondaries which can be incorporated in the ground level materials (carbon-14 is the best known secondary) and their study can tell us about such matters as long-term changes of climate.

Finally, and this is one of the author's current obsessions, there is the possibility of biological changes induced by cosmic rays. There are currently about five cosmic ray secondaries passing through our heads each second, and this flux varies from time to time. It is not inconceivable that in the past (and in the future, too) when the Earth has passed close to a supernova outburst, or the Sun has emitted a singularly large solar flare, the increase in flux has been big enough to cause significant biological change. Perhaps near catastrophic events have been produced in this way. Here is a subject that is well worth further study.

The quest for cosmic rays of very high energy

One reason for studying the highest energy particles has already been given—it is these particles that should be deflected least in space and thus should most easily be traceable back to their sources. There is also the general point that one must always push at the frontier as hard as possible, and the highest particle energy frontier is surely a very worth-while one.

The difficulty with the high energy particles is their rarity. Figure 1 shows the energy 'spectrum', i.e. the number of particles per unit of energy falling on unit area from unit solid angle per second. It can be seen that the spectrum falls off at a ferocious rate. Typical values at the 'top end' are: per unit solid angle for an area of 1 km²

1 per week above 10^{18}eV
1 per year above 10^{19}eV
1 per twenty years above 10^{20}eV

It might be thought that, since a detector (scintillation counter, etc.) of area 1 km² is out of the question, such energies could never be reached, but this is where EAS come into their own. The fact that one particle of high energy can initiate a cascade, and that Coulomb scattering causes the cascade to spread laterally in the atmosphere, means that just a few tens of detectors each, say, 10 square metres in area spread over the square kilometre is enough to detect a single particle of energy 10^{19}eV incident anywhere in that area. Admittedly, some of the information is lost about the particle, such as its mass (whether

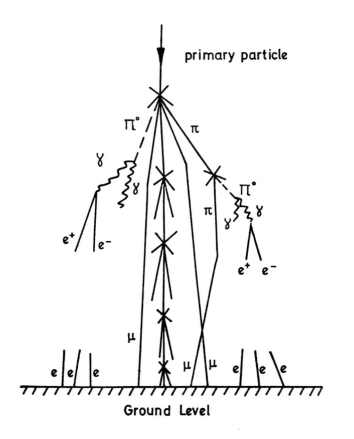

Figure 2. A schematic view of an extensive air shower. The bulk of the particles at ground level are electrons with about 10 per cent muons.

it is a proton, an iron nucleus, a uranium nucleus or whatever...) but a reasonable estimate can be made of its energy
and its direction can be recorded to within a degree or two.
Figure 2 gives a schematic view of a shower, and Plate 1 is an
aerial view of the United Kingdom's EAS array at Haverah Park
near Harrogate.

A prominent feature of the results is a 'kink' in the spectrum
at about 3×10^{15}eV. Although the origin of this feature is not
completely understood, it seems likely that it is because below
this energy the magnetic fields in the Galaxy hold the cosmic
rays for considerable periods (probably ≈ 20 million years) but
above this energy they escape more readily and their 'lifetime' is
shorter. On this model, the bulk of the cosmic rays at these
energies are galactic in origin, i.e. produced in our own Galaxy
rather than being incident on it from outside. If this is true, it
fits in well with the theory put forward in last year's *Yearbook,*
where I showed that the new results in gamma-ray astronomy
could best be understood in terms of cosmic ray protons of
energy in the region of 10^{10}eV having been generated in galactic
sources. The big question now then is—do all the particles right
up through 10^{15}eV and on to 10^{20}eV originate in the same way?

The enigmatic particles above 10^{19}eV

At these extremely high energies if, as seems likely, the
particles are mainly protons, then their deflections in the
galactic magnetic field should be small and their arrival
directions should represent rather faithfully the directions of
their sources. If the particles were of galactic origin, then for
almost any type of source; pulsar, supernova, etc., a higher
intensity would be expected for the direction to the galactic
centre than elsewhere. This is not the case, however; although
the distribution in space becomes increasingly anisotropic as
the energy increases, there is no evidence for a maximum
towards the galactic centre, and the most likely explanation is
that these particles are not, in fact, coming from our own
Galaxy but are incident upon it from outside.

The possibility of an extragalactic origin for the *bulk* of the
cosmic rays was put forward some years ago to account for the

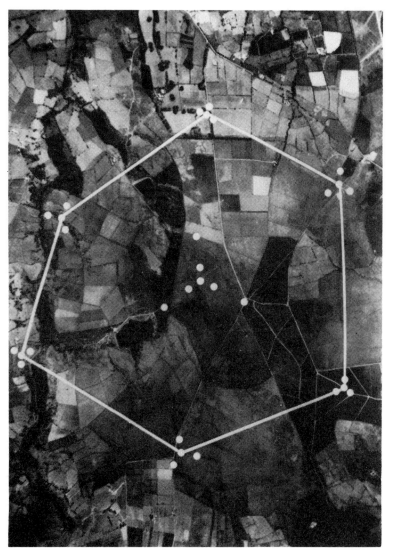

Plate 1. *The Haverah Park Extensive Air Shower Array showing the positions of the major detectors (white circles). The white lines represent the perimeter of the array and enclose an area of 12 km² (courtesy of Prof. J. G. Wilson, West Riding C. C. and Meridian Airmaps Ltd.)*

209

fact that, then, the directional anisotropies seemed to be extremely small. Now, with the new measurements which show quite significant anisotropies, and with the gamma ray data already referred to, a galactic origin seems more likely for the bulk of the particles leaving just the very energetic particles to be so favoured.

It might be thought that there would be no problems with an extragalactic origin for such a small fraction of the flux. After all, the energy content of this fraction is very small, and all that appears to be needed is to have other galaxies, perhaps unusual ones such as Seyfert galaxies or quasars, generating energy spectra which fall off more slowly with increasing energy than does the galactic spectrum. The main problem is that of energy losses incurred by the particles in traversing intergalactic space. Although there may be a little loss by way of interactions with a very tenuous gas in this region (we will come back, later, to the question of gas in between galaxies) the main loss is through interactions with the photons of the 2.7 K relict radiation.

The relict radiation is almost certainly the remains of the radiation generated in the hot big bang, and it corresponds to about 400 photons per cm^3 with an average energy of about 6×10^{-4}eV (i.e. some 2 per cent of the average kinetic energy of molecules in a gas at room temperature). Its effect on protons is as follows. When moving at relativistic velocities the protons 'see' the photons as having much higher energies than the 6×10^{-4}eV they have with respect to the (static) Universe. The increase comes from Einstein's relativity theory, and it can be so big that the photons appear as gamma rays with respect to the protons. Thus, if their energies are more than twice the rest mass of the electron (10^6eV) electron pairs can be generated, energy being taken from the protons and if their energy is even higher (above about 200×10^6eV) pions can be generated and, again, energy is removed from the protons. The result of these interactions is that if protons are generated everywhere in the Universe, their intensity should start to fall at about 10^{18}eV and then fall very rapidly above about 5.10^{19}eV.

Figure 3 shows the spectrum as it would look if, as is likely,

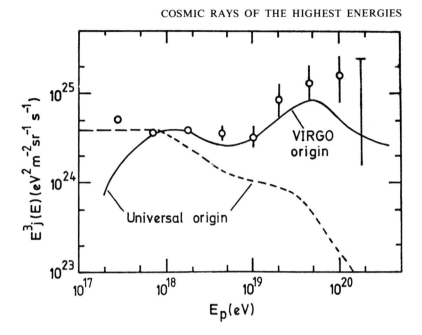

Figure 3. A summary of the energy spectrum of cosmic rays above $10^{17}eV$. The dotted line is what we would expect for an extragalactic origin in the absence of any extragalactic magnetic fields. The full line is the prediction by Wdowczyk and Wolfendale (Nature, Vol. 5730, p. 356, 1979) for the situation where extragalactic magnetic fields cause particle diffusion and virtually all particles seen come from a few energetic galaxies in the Virgo cluster.

the protons have a simple power lower at production. Also shown in the figure is a summary of the world's measurements on the very high energy end of the particle spectrum, from the UK (Haverah Park), USA (Volcano Ranch) and Yakutsk in the USSR. It can be seen immediately that there is no semblance of agreement; where the spectrum should steepen it behaves in just the opposite fashion, i.e. it flattens. In other words, there are many more very energetic particles present above $10^{19}eV$ than would be expected on a simple extragalactic origin model.

The search for an explanation of this paradox has occupied many physicists a lot of time in the last year or two, the present author amongst them. Although the search is by no means over, several ideas have been put forward and one of these—the best yet in the author's view—will be described. It is necessary at this stage to point out that the author has a vested interest in the model—it was introduced by Professor J. Wdowczyk and himself in 1979!

The energy losses on the relict radiation mean that protons starting from 'nearby' sources are at a premium. Essentially this means that their flight times have been so short that they have not had chance to collide with a 2.7 K photon. Protons from our own supercluster of galaxies would do; the 'centre' of the supercluster is near the very energetic galaxy M87 in the Virgo cluster some 20 Mpc away ($1 \text{Mpc} \simeq 3.10^{24}\text{cm}$) i.e. about one three-hundredth of the distance to the 'edge' of the Universe. This has been realized for a long time, but the problem has been to explain why we were not flooded by particles of energy where losses are unimportant (below about 10^{18}eV) from energetic galaxies and clusters further away.

Wdowczyk and I think that the answer comes from the presence of a weak and somewhat tangled magnetic field between galaxies in clusters (and to a lesser extent outside them). There is no direct evidence for this field, but indirect evidence comes from the presence of intergalactic gas. The discovery of this gas is one of the outcomes of contemporary work on cosmic X-rays; X-ray astronomers have found a diffuse component of X-rays coming from clusters which is best explained as due to incandescent gas between the clusters. The presence of such gas had been suspected for a long time, and in fact those who believe in a closed Universe require a 'closure density' which is best achieved if intergalactic gas is widespread.

Much of the gas will be ionized, and the equipartition theorem predicts roughly equal energy in gas motion and the associated magnetic field—the strength of the field is then about 10^{-8} gauss, i.e. a value of about 3.10^{-3} of the interstellar magnetic field and 10^{-7} of the Earth's magnetic field at the surface of the Earth. Although very weak, this field is sufficient when acting over distances of megaparsecs to deflect particles of energy up to 10^{20}ev, and this is just the energy region we are interested in. The only remaining parameter to be defined is the 'cell length' of the magnetic field, i.e. the distance over which its direction is roughly the same; in our model we adopt a value of the order of 0.1 Mpc because this is a few times the overall dimensions of the constituent galaxies. The stage is now set for a reasonable model of extragalactic cosmic rays.

The problem is a straightforward one of diffusion of particles, at least to 10^{20}eV. If we assume that a smallish number of exotic galaxies in each cluster generate the very energetic cosmic rays (such as M87) and that the diffusion coefficient varies with energy as $E^{1/2}$—the dependence which seems to be appropriate for our own Galaxy—then it follows that only particles from the Virgo cluster will arrive at the Earth. Particles from other clusters have not yet diffused far enough to reach us and, indeed, those below about 10^{17} from Virgo have yet to arrive. Energy loss by interaction with the relict radiation is still important, but now with diffusion taking place there is a pile-up of particles and a serious distortion of the energy spectrum will take place. The distortion is just what we require to achieve the measured spectral shape, as can be seen from Figure 3; the model must, therefore, be taken very seriously.

Finally, we return to the problem of directional effects. On this model, well above 10^{20}eV, the detected particles will travel in virtually straight lines from their sources in the Virgo cluster and this should be seen directly. So far, of course, we only have measurements extending to a little above 10^{20}eV but already there is the suggestion that there is some measure of beaming from that direction. Figure 4 shows the latest results from Haverah Park (courtesy of Dr A. A. Watson) and one point from the Yakutsk EAS Array together with our prediction. The agreement is rather remarkable.

Conclusions

Although there can be no absolute certainty about the origin of cosmic rays the most likely situation is as follows. Below about 10^{17}eV galactic sources are mainly responsible, but at

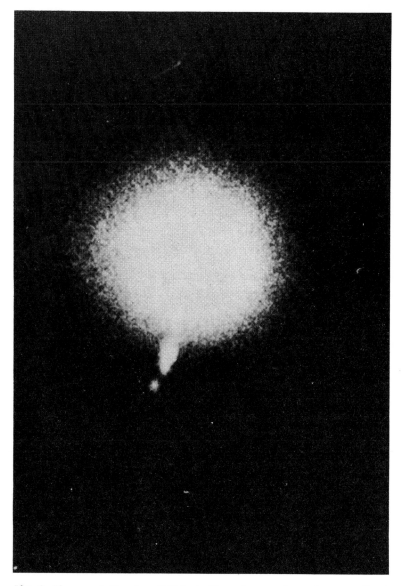

Plate 2. The remarkable galaxy M87 in the Virgo cluster and near the centre of 'our' supercluster. The jet is a likely source of energetic cosmic rays.

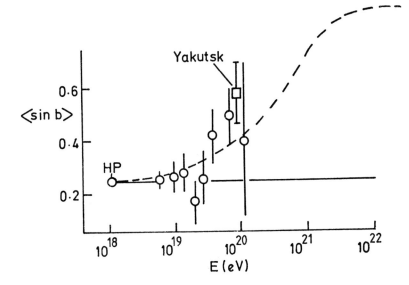

Figure 4. Directional effects for the highest energy cosmic rays. 'b' is the galactic latitude of the particle; the circles are from the Haverah Park experiment and the square is from Yakutsk (marked as such). The full line is expectation for isotropy, i.e. equal numbers from all directions in the sky at all energies. The dotted line is the expectation for the diffusion model of Wdowczyk and Wolfendale.

higher energies extragalactic sources become increasingly important. Of the extragalactic possibilities, the exotic galaxy M87 with its gigantic jet (Plate 2) and similar very energetic objects are most likely. Hopefully, new measurements being made in various countries will confirm these ideas. If not, the hunt will be on again to answer the 70-year old question 'Where do cosmic rays come from?'

The Disappearing Death of Stars

MARTIN COHEN

This will be a story without an ending. There is much we have learned about the lives of stars, but so much more of which we are ignorant. It may be years before we can set into perspective the details of this story but that is the eternal fascination of astronomy—the constant striving to integrate all the facts into the correct framework—to understand fully.

The sunlight filtered through the evergreens by the library windows as I set up the coordinates of the object on the deep photographs of the sky. I was on Mt Hamilton at California's Lick Observatory, making last-minute preparations for an observing spell on the giant 120-inch reflector. The basic programme was to be the continuation of a four-year-long study of young stars. However, at the end of the night, I wanted to squeeze in one or two objects as a pilot to a possible large study. I was working with the US Air Force to identify infrared sources that they had discovered in the course of several short rocket flights designed to explore the infrared sky, at wavelengths that cannot be effectively surveyed from the ground.

I was armed with a list of source coordinates, good to perhaps a couple of arc minutes in each direction. The 'error box' amounted to an area perhaps to a few millimetres square on the sky photographs I was using, large enough to include anywhere from a handful to a hundred star images, depending on where exactly the source lay, relative to the busy plane of our Galaxy. Source 2392 was ready to be located. I set a small paper clip up on each photograph of the region—one taken with a red plate and filter combination, the other with a blue one. The patch of sky on which I concentrated was quite crowded. Quickly I scanned from blue to red and back, studying the photographs through a ten-power eyepiece. One star attracted my attention. At its position on the blue picture, there was nothing, to the limit of the print. On the red was a faint but crisp black dot. I felt I had a potential optical counterpart to the infrared object in the form of this very red star. A few minutes later I had made ten-times enlargements of the two photographs showing the candidate star.

Source 2425 was next. This time the sky was less crowded. No obvious star came up on the red that was absent from the blue picture. However, close to the error box lay a star that was unmistakably brighter on the red than the blue, perhaps by a couple of magnitudes. Again, I took a photograph to serve as a finding chart at the television used to acquire objects at the Cassegrain focus of the 120-inch. I was ready for the night!

It was a good night, with a late-rising moon, although the city lights of San Jose flashed and sparkled below us. The young star study went well, and according to schedule, and I had an opportunity to fit in the two rocket sources. The huge telescope swung around the sky and the massive dome thundered after it, rotating the big slit. The television was turned on, and I looked at the pattern of faint stars it showed, and compared them with the chart for 2392. It was easily identifiable and appeared even brighter on the TV screen, relative to its neighbours, than on the red chart. That instantly told me I probably had the infrared source, for the TV has a phosphor sensitive to near-infrared wavelengths and this star was certainly bright there (at around 8,500 Å). Up on the oscilloscope that represents the spectrum of whatever is in the slit of the ingenious Cassegrain spectrum scanner, a bizarre spectrum was growing before our incredulous eyes.

The left half of the screen (blue to yellow radiation) showed absolutely no signal. At the centre was an abrupt step, then (Figure 1) a strong signal climbing to longer (redder) wavelengths. Altogether the appearance of the spectrum was almost nightmarish! However, it was identifiable as a 'carbon star'. These stars are thought to represent very late phases of stellar evolution, far beyond the long sojourn on the main sequence.

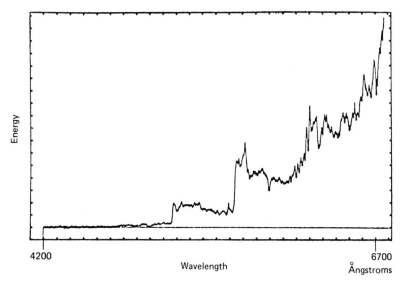

Figure 1. The optical spectrum of source 2392, a carbon star.

Carbon stars are one type of red giant star. Their name derives from their spectra, which are dominated by features due to carbon-bearing molecules (C_2, carbon molecules, and CN, cyanogen). But the usual aspect of source 2392 was the rate at which its energy distribution was increasing through the red region, and into the near-infrared. I was intrigued.

We fed the epoch 1950 coordinates of source 2425 into the computer that resides in the Cassegrain control room, an air-conditioned room (bad for allergies but catering ostensibly to the requirements of the computer's environment) isolated from the cold dome. The machine generated new coordinates, allowing for refraction, precession, and flexure in the long, massive fork mounting. The dome trundled again and we were there. I stood in front of the TV monitor, waiting for the camera inside the telescope to warm up. Faint stars danced against a dark grey sky, scintillating and swimming because of the very low declination and consequently long path through the terrestrial atmosphere. My candidate star was readily located, and I steered it into the slit of the spectrograph.

I watched the spectrum building up before my eyes. This time we had the grating (the dispersive element in the spectrograph that breaks incoming radiation into its different component colours) set to examine the longest wavelengths that we could observe with the instrument (the 8,000-9,000 Å region, beyond the response of our eyes). A spike stood up from the noise almost immediately. It appeared as a finger, pointing straight up. It was soon joined by other, smaller 'fingers', until the spectrum resembled a hand, indicating the heavens (Figure 2). We had caught another bizarre object, clearly the optical counterpart to the infrared source, on the basis of its incredible energy distribution, extremely steeply rising towards the longest wavelengths. But 2425 was not like 2392. This time we were viewing a star that had an oxygen-rich atmosphere, a very late M-type spectrum, dominated by deep absorption bands of titanium oxide. Yet the redness of the starlight was quite unprecedented, among M-stars.

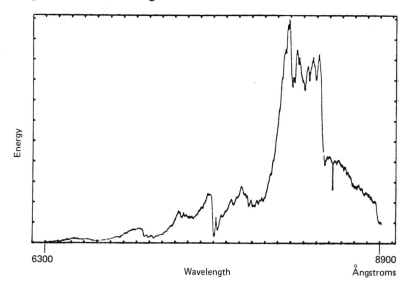

Figure 2. The near-infrared spectrum of source 2425; note the 'hand' near 8300 A caused by deep absorptions of oxides in the stellar atmosphere.

219

Sources 2392 and 2425 share some crucial characteristics. Both are very cool stars, with atmospheres (photospheres) as cool as perhaps 2,500 K (the Sun is about 6,000 K, and Betelgeux, itself an M-star, is about 3,000 K). Both are believed to represent advanced stages in the lives of stars originally not too dissimilar from the Sun in mass. Both are bright infrared sources at the very long wavelengths, with unusually steep optical spectra.

The picture which has emerged of these objects is one in which very cool giant (highly luminous) stars shed a significant portion of their atmospheres as they evolve. This mass loss is initially gaseous, but is able to condense into tiny dust particles as the gas cools while still at high density. Different types of star produce different types of dust grain: the oxygen-rich stars make silicate grains (stomach antacids and talcum powders are silicates), the carbon-rich ones make graphite and silicon carbide (pencil 'leads' and micro-grinding material, respectively). These dust particles are adept at absorbing starlight and, by heating themselves up to some hundreds of degrees, reradiating the energy in the form of infrared radiation. These materials have characteristic infrared spectral features and these have been detected by infrared spectroscopy of cool stars (at around ten microns, some twenty times the wavelength of yellow light). Such particles can be inferred to be around conspicuously bright stars, like Betelgeux and Mira, and it is believed that these stars can spend appreciable periods of time losing matter from their atmospheres and building up their dust shells. But the Air Force objects differ from these bright visible stars in one important respect: their self-made dust shells have apparently become so thick that they effectively prevent most of the star's light from escaping.

We do not yet know whether these thick shells have arisen because some stars always lose mass more rapidly than others, or merely because we are seeing them at a later phase of evolution. Either way our model has two consequences, both in accord with the observations. First, any optical counterparts to the infrared objects should be very faint due to the severe local attenuation of starlight. Secondly, the spectra should be selectively deprived of bluer radiation, this being the nature of the 'extinction' process, that is, the obscuration of light by small dust particles.

There is one deceptively simple issue to raise for all these objects. If they have cool (only hundreds of degrees) infrared-emitting dust shells, why are they detectable at all, since the radiant energy output of thermal objects is a strong function of temperature (as the fourth power)? The answer must be that, although very cool, such stars are extremely large (since energy output is also proportional to radiating surface area). This is even amenable to direct observational checks in the case of a handful of extreme rocket sources. Their faint photographic images are not unresolved points, but rather are tiny extended fuzzy patches, a consequence of the scattering of starlight within enormous circumstellar envelopes of gas and dust.

If we examine what is known about cool red-giant stars we find that, as the stellar temperature is decreased, the proportion of stars known to be variable increases. Variability is characteristic of almost all the coolest stars studied, sometimes periodic, as in the Mira stars, sometimes 'semi-regular', as in the case of Alpha Herculis, for example. At present, too little is known about the optical spectra of the Air Force objects to indicate whether they are Mira-like variables, although some definitely do vary in the infrared. During the next one to two years I hope to observe repeatedly a sample of about twenty of these stars to answer this question. Certainly optical variations occur in some stars, but periodicity cannot yet be established. However, in the case of one notable star, near-infrared observations span over a dozen years now, and it is remarkably well-behaved at around two microns wavelength, varying regularly with a period of almost two years.

One piece of quantitative information my observations yield is an estimate of the total light attenuation of the shells. In particular, I would like to investigate whether this also might vary, indicating changes in the amount of dust present, perhaps due to newly arrived, recently condensed grains at certain phases of the light curve. Additionally, I have seen changes in the spectra indicative of variations in stellar surface temperature, presumably accompanied by alterations in stellar radius. These objects could well be quite actively engaged in pulsations that totally alter their configurations with regard to size, temperature, and dust shell thickness.

Another interesting feature of cool variable stars is the presence, for most of the light cycle, of emission lines that appear throughout the normal absorption line spectra. The hydrogen lines are the principal bright contributors to these emission spectra. Optical spectroscopy has revealed that these lines arise initially deep below the outer atmospheres of the stars. They probably represent shock waves when some mechanism related to the intrinsic variability of the stars drives internal energy mechanically to the surfaces. These emission lines rise higher in the atmospheres and eventually emerge above them. It is possible that these rising shocks are related to the gas shells which are sometimes seen spectroscopically, expanding away from cool variable stars. The nature of the driving force behind the process of mass loss is central to these issues but little of a definitive nature can yet be stated.

During my routine monitoring of one extreme carbon star in 1979, I found that the stellar (absorption) spectrum was apparently almost completely veiled. However, a number of bright emission lines were present that had not been visible in 1977. For several months I watched this weird combination of features but essentially it did not change. I would like to interpret the phenomenon as indicating a new phase of high mass lòss in the star. This freshly ejected material is apparently moving very rapidly (perhaps as fast as 80 km/sec) and is impacting on the previous accumulation of gas and dust, now either static or at least moving very slowly away from the star. Shock waves arise on impact and produce the emission lines as the gas cools behind the shocks. Dust grains also have freshly condensed around the star and serve to veil the starlight. It would be highly exciting if this new stage of evolution, maintained for at least several months, were to mark the end of a previous Mira phase. Perhaps we are about to see this star evolve to the next state, beyond a Mira, and become a planetary nebula!

Many mysteries still remain about cool stars, and some have resisted explanation for several decades. For example, why are the infrared-brightest variable stars also those with the longest periods, and largest amplitudes of variation? How are carbon-rich and oxygen-rich cool stars related in terms of their previous evolution (are they separate phenomena or does a star of one type progress to the other)? It is possible that monitoring of the extreme rocket-infrared sources will offer some clues to the solutions of these and other problems, principally because of their potential for being the latest optically visible phases in the life of an average star.

The New Case Against Extraterrestrial Civilizations

Heaven and Earth are large, yet in the whole of space they are but as a small grain of rice. . . . How unreasonable it would be to suppose that besides the heaven and earth which we can see there are no other heavens and no other earths?—Teg Mu, thirteenth-century Chinese philosopher, translated by Joseph Needham, quoted in NASA SETI report, 1977

Since at least the 19th century it has been considered the height of philosophical naïveness in scientifically sophisticated circles to believe that the Earth is something extra special in the universe—molecular biologist Gunther S. Stent, NY Times Book Reviews, *1974*

The Galaxy is so large that it must *contain other intelligent life-forms, many of them further advanced than ourselves—Duncan Lunan,* Interstellar Contact, *1974*

An attitude which asserts that man is the only intelligent life form in the universe is intolerably arrogant . . . Anyone who holds such an opinion today is . . . an intellectual freak. The odds against intelligent life occurring fairly frequently within our galaxy are impossible ones . . .—Robert K. G. Temple, The Sirius Mystery, *1976*

The source of all this confidence in the existence of many forms of extraterrestrial intelligence (abbreviated ETI) is the near-infinite size of the Universe which, multiplied by any probability greater than zero, will yield a very large number. Even if the odds of 'good' stars, 'good' planets, life beginning, life evolving, and intelligence developing are all tiny, when multiplied together and then multiplied by the total number of opportunities, the answers always seems to give numbers ranging from 'several' to 'several million'.

The formal expression of this intuitive argument is found in 'Drake's Equation', which has the form of

$$N = F_p \times n_e \times f_l \times f_i \times f_c$$

where the factors are either fractions of some event happening, or numbers of chances (numbers of rolls of the dice, metaphorically speaking). Since discoveries in astronomy and biology have indicated that planets *are* not unlikely, that life is *not* unlikely, and that evolution *does* seem to be a directed process, those spokesmen who are confident in the existence of ETI (generally based on philosophical rather than scientific grounds) can find plenty of evidence to support their intuitions.

But wait just a moment. In the excitement over the ETI predictions, and the much-publicized programmes for CETI ('communicating' with ETI) and SETI (a less ambitious 'search' for ETI), many contrary opinions had been overlooked. Many astronomers, biologists, and geologists have been insisting all along that there are good scientific reasons to doubt the validity of Drake's Equation and all its consequences. Ironically, during the triumphant years of SETI in the 1960s and the 1970s, new evidence was being gathered which was chipping away at the popularly accepted 'proofs' for other civilizations in the Galaxy.

These heretics usually worked and wrote (and their articles were usually rejected) singly and alone, but as the evidence grew, so did their boldness in presenting their views. The movement reached 'critical mass' in late 1979 when a special conference was organized at the University of Maryland to discuss the provocative topic, 'Where are they? Implications of our Failure to Observe Extraterrestrials.' Several dozen specialists in various fields gathered at this closed symposium on the weekend of 2-4 November, and put forth their long-simmering doubts.

Blasting the validity of Drake's Equation, conference co-chairman Dr Michael Hart (an astronomer from Trinity University in San Antonio) asserted: 'All of these factors are pure guesswork! There is an uncertainty factor of *millions,* even without the factor of life formation. . .' Hart continued, 'For centuries people have wondered if there is life on other planets. . . Some astronomers have strongly pushed the view that life is common, that intelligent life is common, that high technology abounds in the galaxy . . . That general line of reasoning has proven persuasive but there is a drawback— where is everybody? This has led many of us to doubt the presence of extraterrestrial life.' This question, first expressed as 'Fermi's Paradox', involves the apparent absence of evidence for ETIs both within our own Solar System and also in observations, by radio, optical, and high energy receivers, elsewhere in our galaxy and in other galaxies. 'Why do people reject these conclusions?' Hart wondered. 'I think a distaste for the results make them doubt the argument. People want to believe there is something out there. This has a good place in wishful thinking, not in science. We are alone in the observable universe, and for better or for worse, our fate is in our own hands. . .'

Endorsing the conclusion and underlining the primarily non-rational grounds for the ETI/anti-ETI conflict, radio astronomer Benjamin Zuckerman (who co-directed the OZMA-2 SETI radio search in the mid-1970s) commented that 'there aren't many compelling scientific arguments on either side of this issue, although I think the weight of evidence now does suggest that we have the biggest brains in the galaxy. That's my intuition based on the evidence at hand . . . If I had known then (during OZMA-2) what I know now, I would not have spent my time on the project.'

Sebastian von Hoerner, a leading scientist at the Green Bank National Radio Observatory, added his comments: 'The basic assumption of our previous estimates, that we on Earth are about average and nothing special, seems to lead to a serious contradiction. . . . All that we imagine as our own probable future, large-scale space exploration and colonization, all this should have happened long ago. . . . The whole galaxy should be teeming with life, so obvious that there is no question about it . . .' Writing in the *Journal of the British Interplanetary Society* in 1975, von Hoerner had been more specific: 'Why do we see no interstellar activity, not a single sign of life, among the billions of stars in our own galaxy, among the billions of galaxies visible in our telescopes? This, indeed, is a very serious question. And our ignorance of what to expect is rather unimportant. Like some still untouched natives would have no way of guessing what a city looks like: but if some day they would see one, they would immediately recognize it as something artificial. . . . Somehow we feel we should never expect to be something special, but maybe we are. It cannot be excluded that instead of being average, we are one of the first, and nobody else has developed the ability or had the time to do something being visible to us.. . .'

That argument also has convinced noted physicist Freeman Dyson of the Center for Advanced Studies in Princeton. He reiterated what he had written fifteen years before: 'I feel that if some technological civilization had really ever gotten loose in the Galaxy, the results would be overwhelmingly obvious. Stars would be tamed and rebuilt. Energies from interstellar vehicles would be detected all over the place. The absence of such indications leads me to doubt the presence of extraterrestrial civilizations.' The same line is now also being voiced by Russia's leading astronomer, Dr Yosef Shklovskiy (See Dr Shklovskiy's article in the 1980 *Yearbook*), co-author (with Carl Sagan) of *Intelligent Life in The Universe* (1965). 'Practically, if not absolutely, we are alone in the universe,' he wrote in a Russian philosophical journal in 1976, basing his conclusion on the absence of 'space miracles' which would be the by-products of alien technologies. The following year he clarified his position during a debate in a popular science monthly: 'I do not claim that I have proved our uniqueness in the Universe,' he pointed out, 'but have merely shown that a statement about our *practical* solitude is significantly better justified by specific scientific facts than the traditional common opinion about the multiplicity of civilizations.'

Dr Ronald Bracewell, whose book *The Galactic Club* (1974)

popularized the idea of alien robot probes sent to monitor candidate solar systems for the eventual development of intelligent life, also rejected the tyranny of Drake's Equation at the anti-ETI conference: 'It is perfectly possible that intelligent life, or even life, is virtually unique in the Galaxy.' He made an analogy with the rise of intelligent life on *Earth,* which, despite the many different species which could have given rise to intelligent beings, saw it happen only *once:* 'Why did only one intelligent species arise on Earth? Why no smart prairie dogs, raccoons, otters, baboons, kangaroos, and so forth? They could not, because homo sapiens arose in one spot and then moved out and occupied the entire planet, pre-empting all other areas where other intelligences might (if left alone) have arisen... Estimates do not eliminate the prospect that the first galactic species whose biological evolution brings technology within reach may spread over its planet, through its planetary system, and continue on into the galaxy to preempt or incorporate alien cultures should such be encountered.'

Nor are astronomers alone in this scepticism. As pointed out by Dr Frank J. Tipler in an unpublished paper in 1979, 'Most leading experts in evolutionary biology, such as Dobzhansky, Simpson, Francois, Ayala *et al,* and E. Mayer, contend that the Earth is probably unique in harboring intelligence, at least amongst the planets of our galaxy.'

To summarize this argument: all deductions *a priori,* that is, based on *guesses* as to the actual values of the factors of Drake's Equation, are rejected; additionally, Drake's Equation itself is rejected as incomplete and simplistic. *A posteriori,* these anti-ETI advocates set up a reasonable model which they suggest would be the course of development for any advanced spacefaring civilization—and that model results in effects which are not observed, neither 'locally' nor in the Galaxy at large. Consequently, either the model is wrong (which is certainly possible), or there do not exist any ETI civilizations.

A key topic of dispute in the model is the possibility of interstellar travel. Flitting from star to star within a few days, at 'warp speed' (à la *Star Trek*) is not necessary; the anti-ETI school would be quite satisfied with centuries-long voyages, or even with robot craft which crawl from star to star over a period of thousands of years. The loudest advocates of the existence of ETI in the astronomy community do not advertise (but do not deny, either) their assessment that any interstellar travel is highly improbable if not absolutely impossible: else they too would be faced with the nagging question, 'Where are they?' As Isaac Asimov put it in his book *Extraterrestrial Civilizations* (1979), 'Once we allow the practicality of easy interstellar travel, we are forced to speculate that Earth is being visited or has been visited, is being helped or at least left alone by a Federation of benevolent civilizations. Well, perhaps, but none of it sounds compelling. It seems safer to assume that interstellar travel is not easy or practical. . .'

Disregarding for a moment that claim that some UFO reports are based on extraterrestrial spaceships, there is definitely a loophole in this anti-ETI argument. In an article in *Science* magazine in 1977, Caltech astronomers T. Kuiper and M. Morris accept the premise that ETI civilizations, if they exist, will achieve interstellar flight and thus will already have arrived in the neighbourhood of Earth. Earlier, Hart argued that since we do not *see* such visitors, they do not exist. But Kuiper and Morris disagree: 'The implications that Hart draws differ from ours,' they point out, 'elaborating on several entirely selfish motives which might keep ETI visitors from overt contact. First, their stage of civilization may not even require planetary bases and they could be happily ensconsed in the asteroid belt; second, Earth's biology could be incompatible with them; lastly, 'the possibilities that we are being ignored, avoided, or discretely watched are logically possible.'

Such arguments would go a long way towards vitiating the oft-heard argument against the reality of UFOs: 'If they really came all the way across the Galaxy, why haven't they made official contact?' Both the article by Kuiper and Morris, and speculations by UFO authors such as Stanton Friedman, have already provided quite plausible scenarios for such a course of (in)action. There are more logical arguments than that one against the ETI nature of UFO reports.

And this reasoning does not eliminate the second leg of the

anti-ETI arguments: why don't we see their activities across the whole Galaxy? Astronomers do know that there must be several supernovæ per century going off in our Galaxy which are unobserved because of blocking clouds of dust—but the kinds of energies which a trans-galactic civilization would give off, in their industrial, transportation, and military activities, should be detectable and recognizable—except we have not recognized them. Something is wrong with the assumptions— or the galaxy is still 'wild'. Perhaps it will be up to our descendants to 'tame' it.

The new-found confidence of these anti-ETI advocates is based in part on their newly-discovered numbers, and in part on hard new evidence. Many of the arguments advanced in the 1960s for the ubiquity of planetary systems have been demolished—claims that extrasolar planets have been detected by 'star wobbles' or by the slow spin of the primary stars have now been abandoned as 'instrumental error' or missing factors. The suggestion that our Sun could be among the oldest of the 'heavy metallic' stars, and thus one of the first to be possible as a habitat for life, has been disproven—there are entirely Sun-like stars many billions of years older than our Sun. New theories of planetary formation have uncovered many random factors which tend to decrease the probability that planets, and especially rocky planets, will form frequently, anywhere.

And once formed—even granting the still problematic spontaneous formation of living cells—the survival of a life-bearing planet is now considered unlikely. Computer simulations have shown, for example, that in Earth's history the planet twice teetered on the brink of total ecological disaster, once when it nearly saw its oceans freeze solid, and another time a thousand million years later, when they nearly boiled away. Different sized stars than the Sun have different temperature histories, so that there are *no* zones near them in which the temperature is continuously habitable for several billion years; different sized planets than the Earth have different rates of atmospheric evolution which also fail to compensate for stellar heating variations, leading to planet-wide sterilization. Additionally, astronomer Gerrit Verschuur has recently called attention to the fatal effects of 'postulated cosmic hazards to life, which include the effect of supernovæ, sunspot cycles, passage through interstellar dust clouds, and being located in regions of the Galaxy where ambient cosmic rays fluxes may be higher than those we are used to. . .', all of which factors would inexorably winnow out the number of planets on which life did manage to form. Additional factors, such as the presence of a nearby large moon, and the coincidental values of the volume of surface water and the capacity of Earth's ocean basins, have not yet been analyzed in detail—but they, too, suggest more and more that it is the life-bearing Earth which is a freak in the galaxy, not the dead worlds of Venus, Mars, and the Moon.'

This new model of our planet has several chilling implications. In analogy, perhaps our current position is that of a drunk who has staggered across a superhighway, surviving collisions with speeding vehicles only by random chance. Safely across, he might imagine that it had been inevitable—until he notices the roadway littered with the bodies of less-lucky drunkards! So far, we have not had a large enough sample size to tell if we were lucky—we haven't seen enough other planets which did *not* make it. However, taking the analogy a step further, it would be foolish for the drunk to imagine he was safely across the highway, since another fleet of sixteen-wheelers could be speeding down the lane he is now standing in. In the same way, *if* we are here *only* because we have survived a long series of near-fatal accidents, there is no reason to suppose that there are not more such planetary disasters speeding down on us, disasters which could yet wipe out the Earth. Luckily, mankind is rapidly approaching the stage of technological development in which active measures could be taken to counteract such disasters.

Arguments against this anti-ETI point of view, and its chilling implications, remain primarily intuitive and philosophical. In criticizing Tipler's paper, pro-ETI advocate Carl Sagan wrote that 'the assumptions have led to a conclusion which is at least highly implausible; therefore, there is probably a flaw in the assumptions'—but the conclusions are only 'highly implausible' under another set of assumptions, those to which Sagan prefers to cling. 'In this debate,' Sagan continues, 'are extremely strong echoes of the confrontation between Copernican and Ptolemaic world views.' And thus the argument continues, based on past analogies, on intuition—and only occasionally on the few facts which are yet available.

PART THREE

Miscellaneous

Some Interesting Telescopic Variable Stars

Star	R.A. h	m	Dec. °		Mag. range	Period, days	Remarks
R. Andromedæ	0	22	+38	18	6.1–14.9	409	
W Andromedæ	2	14	+44	4	6.7–14.5	397	
Theta Apodis	14	00	−76	33	6.4– 8.6	119	Semi-regular.
R Aquilæ	19	4	+ 8	9	5.7–12.0	300	
R Arietis	2	13	+24	50	7.5–13.7	189	
R Aræ	16	35	−56	54	5.9–6.9	4	Algol type.
R Aurigæ	5	13	+53	32	6.7–13.7	459	
R Boötis	14	35	+26	57	6.7–12.8	223	
Eta Carinæ	10	43	−59	25	−0.8– 7.9	—	Unique erratic variable.
I Carinæ	09	43	−62	34	3.9–10.0	381	
R Cassiopeiæ	23	56	+51	6	5.5–13.0	431	
T Cassiopeiæ	0	20	+55	31	7.3–12.4	445	
X Centauri	11	46	−41	28	7.0–13.9	315	
T Centauri	13	38	−33	21	5.5– 9.0	91	Semi-regular.
T Cephei	21	9	+68	17	5.4–11.0	390	
R Crucis	12	20	−61	21	6.9– 8.0	5	Cepheid.
Omicron Ceti	2	17	− 3	12	2.0–10.1	331	Mira.
R Coronæ Borealis	15	46	+28	18	5.8–14.8	—	Irregular
W Coronæ Borealis	16	16	+37	55	7.8–14.3	238	
R Cygni	19	35	+50	5	6.5–14.2	426	
U Cygni	20	18	+47	44	6.7–11.4	465	
W Cygni	21	34	+45	9	5.0– 7.6	131	
SS Cygni	21	41	+43	21	8.2–12.1	—	Irregular.
Chi Cygni	19	49	+32	47	3.3–14.2	407	Near Eta.
Beta Doradûs	05	33	−62	31	4.5– 5.7	9	Cepheid.
R Draconis	16	32	+66	52	6.9–13.0	246	
R Geminorum	7	4	+22	47	6.0–14.0	370	
U Geminorum	7	52	+22	8	8.8–14.4	—	Irregular.
R Gruis	21	45	−47	09	7.4–14.9	333	
S Gruis	22	23	−48	41	6.0–15.0	410	
S Herculis	16	50	+15	2	7.0–13.8	307	
U Herculis	16	23	+19	0	7.0–13.4	406	
R Hydræ	13	27	−23	1	4.0–10.0	386	
R Leonis	9	45	+11	40	5.4–10.5	313	Near 18, 19.
X Leonis	9	48	+12	7	12.0–15.1	—	Irregular (U Gem type).
R Leporis	4	57	−14	53	5.9–10.5	432	'Crimson star.'
R Lyncis	6	57	+55	24	7.2–14.0	379	
W Lyræ	18	13	+36	39	7.9–13.0	196	

Star	R.A.		Dec.			Period	Remarks
	h	m	°	′	Mag. range	days	
T Normæ	15	40	−54	50	6.2–13.4	293	
HR Delphini	20	40	+18	58	3.6–?	–	Nova, 1967.
S Octantis	17	46	−85	48	7.4–14.0	259	
U Orionis	5	53	+20	10	5.3–12.6	372	
Kappa Pavonis	18	51	−67	18	4.0– 5.5	9	Cepheid.
R Pegasi	23	4	+10	16	7.1–13.8	378	
S Persei	2	19	+58	22	7.9–11.1	810	Semi-regular.
R Sculptoris	01	24	−32	48	5.8– 7.7	363	Semi-regular.
R Phœnicis	23	53	−50	05	7.5–14.4	268	
Zeta Phœnicis	01	06	−55	31	3.6– 4.1	1	Algol type.
R Pictoris	04	44	−49	20	6.7–10.0	171	Semi-regular.
L² Puppis	07	12	−44	33	2.6– 6.0	141	Semi-regular.
Z Puppis	07	30	−20	33	7.2–14.6	510	
T Pyxidis	09	02	−32	11	7.0–14.0	–	Recurrent nova (1920, 1944)
R Scuti	18	45	− 5	46	5.0– 8.4	144	
R Serpentis	15	48	+15	17	5.7–14.4	357	
SU Tauri	5	46	+19	3	9.2–16.0	–	Irregular (R CrB type).
R Ursæ Majoris	10	41	+69	2	6.7–13.4	302	
S Ursæ Majoris	12	42	+61	22	7.4–12.3	226	
T Ursæ Majoris	12	34	+59	46	6.6–13.4	257	
S Virginis	13	30	−6	56	6.3–13.2	380	
R Vulpeculæ	21	2	+23	38	8.1–12.6	137	

Note: Unless otherwise stated, all these variables are of the Mira type.

Some Interesting Double Stars

The pairs listed below are well-known objects, and all the primaries are easily visible with the naked eye, so that right ascensions and declinations are not given. Most can be seen with a 3-inch refractor, and all with a 4-inch under good conditions, while quite a number can be separated with smaller telescopes, and a few (such as Alpha Capricorni) with the naked eye. Yet other pairs, such as Mizar-Alcor in Ursa Major and Theta Tauri in the Hyades, are regarded as too wide to be regarded as bona-fide doubles!

Name	Magnitudes	Separation″	Position angle, deg.	Remarks
Gamma Andromedæ	3.0, 5.0	9.8	060	Yellow, blue. B is again double (0″.4) but needs a larger telescope.
Zeta Aquarii	4.4, 4.6	2.6	291	Becoming more difficult.
Gamma Arietis	4.2, 4.4	8	000	Very easy.
Theta Aurigæ	2.7, 7.2	3	330	Stiff test for 3 in. OG
Delta Boötis	3.2, 7.4	105	079	Fixed.
Epsilon Boötis	3.0, 6.3	2.8	340	Yellow, blue. Fine pair.
Kappa Boötis	5.1, 7.2	13	237	Easy.

SOME INTERESTING DOUBLE STARS

Name	Magnitudes	Separation"	Position angle, deg.	Remarks
Zeta Cancri	5.6, 6.1	5.6	082	
Iota Cancri	4.4, 6.5	31	307	Easy. Yellow, blue.
Alpha Canum Venat.	3.2, 5.7	20	228	Yellowish, bluish. Easy.
Alpha Capricorni	3.3, 4.2	376	291	Naked-eye pair. Alpha again double.
Eta Cassiopeiæ	3.7, 7.4	11	298	Creamy, bluish. Easy.
Beta Cephei	3.3, 8.0	14	250	
Delta Cephei	var, 7.5	41	192	Very easy.
Alpha Centauri	0.0, 1.7			Binary; period 80 years. Very easy.
Xi Cephei	4.7, 6.5	6	270	Reasonably easy.
Gamma Ceti	3.7, 6.2	3	300	Not too easy.
Alpha Circini	3.4, 8.8	15.8	235	PA, slowly decreasing.
Zeta Coronæ Borealis	4.0, 4.9	6.3	304	
Delta Corvi	3.0, 8.5	24	212	
Alpha Crucis	1.6, 2.1	4.7	114	Third star in low-power field.
Gamma Crucis	1.6, 6.7	111	212	Wide optical pair.
Beta Cygni	3.0, 5.3	35	055	Yellow, green. Glorious.
61 Cygni	5.3, 5.9	25	150	
Gamma Delphini	4.0, 5.0	10	265	Yellow, greenish. Easy.
Nu Draconis	4.6, 4.6	62	312	Naked-eye pair.
Alpha Geminorum	2.0, 2.8	2	151	Castor. Becoming easier.
Delta Geminorum	3.2, 8.2	6.5	120	
Alpha Herculis	var, 6.1	4.5	110	Red, green.
Delta Herculis	3.0, 7.5	11	208	Optical double.
Zeta Herculis	3.0, 6.5	1.4	300	Fine, rapid binary.
Gamma Leonis	2.6, 3.8	4.3	121	Binary; period 400 years
Alpha Lyræ	0.0, 10.5	60	180	Vega. Optical; B faint.
Epsilon Lyræ	4.6, 6.3	3	005	Quadruple. Both pairs
	4.9, 5.2	2.3	111	separable in 3 in. OG
Zeta Lyræ	4.2, 5.5	44	150	Fixed. Easy double.
Beta Orionis	0.1, 6.7	9.5	205	Rigel. Can be split with 3 in.
Iota Orionis	3.2, 7.3	11	140	
Theta Orionis	6.0, 7.0			The famous Trapezium in
	7.5, 8.0			M.42
Sigma Orionis	4.0, 7.0	11.1	236	Quadruple. D is rather
		12.9	085	faint in small apertures.
Zeta Orionis	1.9, 5.0	3	160	
Eta Persei	4.0, 8.5	28.5	300	Yellow, bluish.
Beta Phoenicis	4.1, 4.1	1.3	352	Slow binary.
Beta Piscis Austrini	4.4, 7.9	30.4	172	Optical pair. Fixed.
Alpha Piscium	4.3, 5.3	1.9	291	
Kappa Puppis	4.5, 4.6	9.8	318	Again double.
Alpha Scorpii	0.9, 6.8	3	275	Antares, Red. green.
Nu Scorpii	4.2, 6.5	42	336	
Theta Serpentis	4.1, 4.1	23	103	Very easy.
Alpha Tauri	0.8, 11.2	130	032	Aldebaran. Wide, but B is very faint in small telescopes.
Beta Tucanæ	4.5, 4.5	27.1	170	Both components again double.
Zeta Ursæ Majoris	2.3, 4.2	14.5	150	Mizar. Very easy. Naked eye pair with Alcor.
Alpha Ursæ Minoris	2.0, 9.0	18.3	217	Polaris. Can be seen with 3 in.
Gamma Virginis	3.6, 3.7	4.8	305	Binary; period 180 yrs. Closing.
Theta Virginis	4.0, 9.0	7	340	Not too easy.
Gamma Volantis	3.9, 5.8	13.8	299	Very slow binary.

Some Interesting Nebulæ and Clusters

Object	R.A.		Dec.		Remarks
	h	m	°		
M.31 Andromedæ	00	40.7	+41	05	Great Galaxy, visible to naked eye.
H.VIII 78 Cassiopeiæ	00	41.3	+61	36	Fine cluster, between Gamma and Kappa Cassiopeiæ.
M.33 Trianguli	01	31.8	+30	28	Spiral. Difficult with small apertures.
H.VI 33 4 Persei	02	18.3	+56	59	Double cluster; Sword-handle.
△142 Doradûs	05	39.1	−69	09	Looped nebula round 30 Doradûs. Naked-eye. In Large Cloud of Magellan.
M.1 Tauri	05	32.3	+22	00	Crab Nebula, near Zeta Tauri.
M.42 Orionis	05	33.4	−05	24	Great Nebula. Contains the famous Trapezium, Theta Orionis.
M.35 Geminorum	06	06.5	+24	21	Open cluster near Eta Geminorum.
H.VII 2 Monocerotis	06	30.7	+04	53	Open cluster, just visible to naked eye.
M.41 Canis Majoris	06	45.5	−20	42	Open cluster, just visible to naked eye.
M.47 Puppis	07	34.3	−14	22	Mag. 5,2. Loose cluster.
H.IV 64 Puppis	07	39.6	−18	05	Bright planetary in rich neighbourhood.
M.46 Puppis	07	39.5	−14	42	Open cluster.
M.44 Cancri	08	38	+20	07	Præsepe. Open cluster near Delta Cancri. Visible to naked eye.
M.97 Ursæ Majoris	11	12.6	+55	13	Owl Nebula, diameter 3'. Planetary.
Kappa Crucis	12	50.7	−60	05	"Jewel Box"; open cluster, with stars of contrasting colours.
M.3 Can. Ven.	13	40.6	+28	34	Bright globular.
Omega Centauri	13	23.7	−47	03	Finest of all globulars. Easy with naked eye.
M.80 Scorpii	16	14.9	−22	53	Globular, between Antares and Beta Scorpionis.
M.4 Scorpii	16	21.5	−26	26	Open cluster close to Antares.
M.13 Herculis	16	40	+36	31	Globular. Just visible to naked eye.
M.92 Herculis	17	16.1	+43	11	Globular. Between Iota and Eta Herculis.
M.6 Scorpii	17	36.8	−32	11	Open cluster; naked-eye.
M.7 Scorpii	17	50.6	−34	48	Very bright open cluster; naked eye.
M.23 Sagittarii	17	54.8	−19	01	Open cluster nearly 50' in diameter.
H.IV 37 Draconis	17	58.6	+66	38	Bright Planetary.
M.8 Sagittarii	18	01.4	−24	23	Lagoon Nebula. Gaseous. Just visible with naked eye.
NGC 6572 Ophiuchi	18	10.9	+06	50	Bright planetary, between Beta Ophiuchi and Zeta Aquilæ.
M.17 Sagittarii	18	18.8	−16	12	Omega Nebula. Gaseous. Large and bright.
M.11 Scuti	18	49.0	−06	19	Wild Duck. Bright open cluster.
'M.57 Lyræ	18	52.6	+32	59	Ring Nebula. Brightest of planetaries.
M.27 Vulpeculæ	19	58.1	+22	37	Dumb-bell Nebula, near Gamma Sagittæ.
H.IV 1 Aquarii	21	02.1	−11	31	Bright planetary near Nu Aquarii.
M.15 Pegasi	21	28.3	+12	01	Bright globular, near Epsilon Pegasi.
M.39 Cygni	21	31.0	+48	17	Open cluster between Deneb and Alpha Lacertæ. Well seen with low powers.

Our Contributors

JOHN MASON is a microcomputer designer, and researcher in high-voltage discharges and lightning. He is a graduate of University College, London. He is a member of the British Astronomical Association, specializing in observations of comets and meteors and in the application of electronics to astronomy.

J. LEONARD CULHANE is one of Britain's leading experts in artificial satellite technology. He is based at the Mullard Space Science Laboratory (University College London) at Holmbury St. Mary, and has been deeply involved in all the problems of Britain's latest satellites, including Ariel VI.

GORDON E. TAYLOR, of the Nautical Almanac Office, Royal Greenwich Observatory, Herstmonceux, has specialized in all branches of mathematical astronomy, and has developed the new technique of measuring the apparent diameters of small solar system bodies by occultations. He is a Past President of the British Astronomical Association, and is now Director of its computing section and Editor of its *Handbook*.

DR GARRY E. HUNT, Head of the Laboratory for Planetary Atmospheres, Department of Physics and Astronomy, University College London, is one of the world's leading experts in planetary research and is closely concerned with current space probe projects, particularly Viking and Voyager. In addition to his technical work, he is also well known for his television and radio broadcasts in Britain, the USA and Canada, and popular lectures.

239

54284

DAVID A. ALLEN is a Cambridge graduate; he then spent some years at the Royal Greenwich Observatory, and is now at Siding Spring Observatory in Australia. He has concentrated upon infrared astronomy, and has been responsible for important advances in this field. He is, of course, one of our most regular and welcome *Yearbook* authors.

ARNOLD W. WOLFENDALE, F.R.S., is Professor of Physics at the University of Durham. His research work extends over many fields of physics and astronomy, and in particular he has been undertaking fundamental research in connection with high-energy radiations, including gamma-rays and cosmic rays.

MARTIN COHEN is a Cambridge graduate who, like Dr David Allen, has been concerned largely with infrared research; latterly he has been based in the United States, using the powerful equipment there to carry out pioneer research. He, too, is one of our regular and valued contributors.

JAMES E. OBERG, of the National Aeronautics and Space Administration, has been deeply concerned with space-probe and satellite research. He has also carried out investigations into aerial phenomena and into the possibilities of extraterrestrial life.